MW00835236

W. Birkmayer · P. Riederer

Understanding the Neurotransmitters: Key to the Workings of the Brain

Translated by Karl Blau

Springer-Verlag Wien GmbH

Prof. Dr. Walther Birkmayer
Birkmayer-Institut für Parkinsontherapie, Vienna, Austria

Prof. Dr. Peter Riederer
Clinical Neurochemistry, Department of Psychiatry,
University of Würzburg, Federal Republic of Germany

Translation of
„Neurotransmitter und menschliches Verhalten"
Wien-New York: Springer-Verlag 1986

Originally published by Springer Vienna in 1989
Softcover reprint of the hardcover 1st edition 1989

With 14 Figures

Cover: Abstract figure of brain function. Single paths mimic the function of certain neurotransmitter systems. Crossing points symbolize their interference. The chemical balance within the brain is reflected by mental harmony (and vice versa)
(Design: Dr. Paul Kruzik, Vienna; Drawing: Wolfgang Rieder, Vienna)

ISBN 978-3-211-82100-8 ISBN 978-3-7091-3451-1 (eBook)
DOI 10.1007/978-3-7091-3451-1

We dedicate this book to our wives
Anny Birkmayer (Besler)
and Inge Riederer (Winkelmayer)
with grateful thanks for their collaboration
and interest in our work over so many years

Foreword

Walther Birkmayer, one of the co-discoverers of the neurochemical causes of Parkinson's disease and of its treatment by neurotransmitter-replacement therapy, and his former collaborator and colleague Peter Riederer have succeeded in this book in giving us an outline and overview of current knowledge of the biochemistry of the brain and synaptic transmission between nerve cells, together with an appreciation of how these factors relate to recent findings in human behavioural science, an area in which both authors have made substantial contributions.

Ethology is the systematic study of behaviour, and its importance was recognized by the award of the Nobel Prize for Medicine to its founders and principal exponents Konrad Lorenz, Niko Tinbergen and Karl von Frisch. Human behavioural science extends their methods to human beings, and we owe the application of these methods in the human field to among others Irenaeus von Eibl-Eibesfeldt as far as normal human behaviour is concerned, and to Dieter von Ploog for the study of abnormal behaviour. These workers have played a decisive part in helping us to understand the biological factors that govern human behaviour. Fundamental to all these studies is the way we humans feel, and thanks to our ability to express our feelings, thoughts and emotions in words, we can accurately describe these feelings, thoughts and emotions, not only in normal circumstances but also when we believe them to be disturbed.

For these reasons the chapters on Parkinson's disease, on clinical depression, and on the autonomic nervous system and its relationship to emotional states, are of crucial importance, and the authors have dealt with these in a masterly fashion, describing in some detail the problems raised by the modern fashion of regarding these disorders from a psychosomatic or holistic viewpoint. Intimately linked to this treatment are the problems involving pain and sleep. In the chapters on neurotic developments and personality defect disorders the authors are at pains to emphasise that the neurobiological approach is basic to the proper understanding of both normal and of disordered mental health, at the same time bearing the psychosocial aspects in mind.

It is precisely in those chapters that the book so clearly shows the need for both approaches when tracking down these phenomena. However, the neurobiological approach has the advantage of resting on a definite scientific theory, logical positivism, or in more modern parlance, the critical rationalism of Karl Popper; psychosocial "sciences" on the other hand suffer from the lack of a unifying scientific framework or theoretical basis, because their individual component techniques cannot immediately be applied, or at least not without calling their scientific foundations in question.

This is a book of tremendous knowledge, imbued with the spirit of enquiry, which presents much important research in a very clear and concise fashion and also draws it together into a coherent whole. It demonstrates yet again how urgently the psychosocial sciences need a similarly adequate scientific basis in order to match this very fundamental book. Perhaps "Evolutionary Epistemology", of which clear traces may be found throughout this important work, may show us the way forward.

Professor W. Pöldinger
Medical Director of the University Psychiatric Clinic,
Basel University, Switzerland

Preface

The present book is the comprehensive outcome of more than forty years of clinical observations. It describes not only the methods of treatment but also the theoretical framework underlying these methods of treatment which grew out of those observations. The aim has always been to visualize the human being as a whole, both in himself and within his environment. It all began when Professor Walther Birkmayer was working at the "Wiener Hirnverletzten-Lazarett", the Hospital for the Brain-damaged in Vienna, between 1942 and 1945. This phase did not lead to new insights into damage to the cerebral cortex as much as fresh observations concerning the instinctual region of the brainstem. These observations, which were extensively described in a scientific monograph, led to the realisation that an injury does not necessarily have to cause negative symptoms, but that such lesions may disturb the balance between the various effector substances, resulting in either negative or positive symptoms. In harmony with this view was our demonstration, in numerous patients with injuries to the brainstem, that stressing such patients with adrenalin or with insulin (following suggestions by Professors K. Eppinger and F. Hoff) led to the eliciting of various somewhat paradoxical forms of response. We labelled such inappropriate responses "vegetative ataxia", and took the view that defects on feedback regulation were responsible for such errors of coordination within the autonomic system. From a clinical standpoint it became clear that various signs of clinical deficiencies such as intolerance to alcohol, the premature decline of mental abilities, hypersensitivity to changes in the weather, depressive manifestations and emotional discharges or apathetic reactions were implicated; in other words that we could relate the patient's overall behaviour to the severity of the damage to the brainstem.

What was missing, however, was any objective evidence. The discovery of the biogenic amines and of the mechanisms for their synthesis, so ably carried out by research workers such as Peter Holtz, Hermann Blaschko and Marthe Vogt, made it possible for us to carry out experimental investigations into possible correlations between "models of behaviour" and the function of these transmitter substances. For example, Brodie was able to show that reserpine produced a depletion of certain transmitters from nerve cells. This pharmacological effect could be clinically related to the physiological effect of a lowering of the blood pressure and the psychic effect of depression. The decisive breakthrough which provided crucial evidence for the hypothesis that the behaviour of man and beast is directed by neurotransmitters, was provided by Arvid Carlsson. He and his colleagues discovered in animal experiments that there is a connection between depletion of the biogenic amine dopamine and motor deficiencies. Administration of dopa, the precursor of dopamine, lifted this motor deficit. Thus it could be demonstrated that a specific effect caused by a specific chemical substance, reserpine, could be counteracted by another specific chemical substance, namely L-dopa. This experiment is regarded as one of the crucial events of neurology.

When a range of suggestions based on clinical observations is confirmed by anatomical, pathological or biochemical findings we may justifiably speak of a genuine scientific quantum leap. We have described this as the "evolution of a cognitive conjunction", or in plain language that it takes the combination of clinical observation and fundamental research to produce advances in science. Such a conjunction occurred in 1971, when the biochemist Peter Riederer joined the Ludwig Boltzmann Institute for Neurochemistry. This book and the body of work it describes, are evidence of the success of this fruitful collaboration. Naturally this body of work is not the final version: we are very conscious that it represents only a staging post on the way. Nevertheless even at this stage we already have the ability to design more rational methods of treatment. We are well aware that there must be many transmitter substances that have yet to be discovered, and that our present state of knowledge does not yet allow us to appreciate or to evaluate the precise interrelationships of the various nervous systems. This book deals with what we might call islands of knowledge, but it will call for a lot more research before we have amassed enough facts to integrate these islands into an exact and coherent scientific archipelago, one that is firmly based on evolutionary epistemology and that provides us with the insights which are the stimuli to further research.

We now have the ability of both measuring and adjusting chemical and clinical parameters to a considerable degree: in other words, achieving a balance between the various neurotransmitters is not only a basic need of human behaviour: re-establishing this balance must also be the objective of any specific treatment.

Sigmund Freud, as early as during the first world-war, predicted that "... biochemistry will discover substances that will confirm or refute our hypotheses." Well of course it is hardly up to biochemistry to refute psychoanalytic hypotheses: for example the Oedipus complex cannot be biochemically proved or disproved. On the other hand we *can* recognise that in someone with oedipal tendencies there may be a release of activating neutrotransmitters such as noradrenaline, leading to aggressive tendencies, increased drive and a potentiation of cultural, spiritual and indeed financial abilities, but alternatively there may be a resignation on the part of the son, where a deadening of the emotions may be paralleled by a predominance of parasympathetic activity e.g. serotonin.

We have tried in this detailed preface to introduce readers to an area of enquiry in which knowledge and insight will give them a better understanding of the sources not only of their own behaviour but also that of their fellow human beings.

Walther Birkmayer and Peter Riederer

Acknowledgements

It gives us great pleasure, at this point, to express our warmest thanks to Dipl.-Ing. Dr. P. Kruzik for his contribution to the design of the illustrative figures. We are most grateful to Springer-Verlag, Vienna, for the outstanding production of this book. Our grateful thanks are also due to Dr. E. Handerek and Mrs. I. Riederer for the careful preparation of the manuscript. We are grateful to Dr. Karl Blau who not only translated the words of our book but also managed to convey something of what lies between the lines.

Contents

Generic and trade names

Generic name	Trade names		
	Federal Republic of Germany	United Kingdom	United States of America
Amantadine	Contenton PK-Merz	Symmetrel	Symmetrel
Amiloride	Moduretic	Amilco Berkamil Frumil Hypertane Kalten Lasoride Midamor Moduretic Normetic	Moduretic
Amitriptyline	Tryptizol Limbitrol	Domical Elavil Lentizol Limbitrol Triptafen Tryptizol	Elavil Endep Limbitrol
Amphetamine	Amphetamin	Dexedrine	
Baclofen	Lioresal	Lioresal	Lioresal
Benserazide (+ L-Dopa)	Madopar	Madopar	
Benzhexol	Artane	Artane Bentex Broflex	Artane
Biperidine	Akineton		Akineton
Bornaprin	Sormodren		

Generic name	Trade names		
	Federal Republic of Germany	United Kingdom	United States of America
Bromazepam	Lexotanil	Lexotan	
Bromocriptine	Parlodel Pravidel Umprel	Parlodel	Parlodel
Carbidopa (+ L-Dopa)	Sinemet Nacom	Sinemet	Sinemet
Cerebrolysin	Cerebrolysin		
Chlorprothixen	Truxal	Taractan*	Taractan
Clobazam	Frisium	Frisium	
Clomipramine	Anafranil	Anafranil	
Clonazepam	Rivotril	Rivotril	Clonopin
Clopenthixol	Cisordinol		
Cortisone	Cortison	Cortelan Cortistab Cortisyl	
Dexamphetamine	Amphetamin	Dexedrine	
Diazepam	Valium	Alupram Atensine Diazemuls Evacalm Solis Stesolid Tensium Valium	Valium
Dibenzipine	Noveril		
Dihydro-ergotamine	Dihydergot	Dihydergot	Plexonal Scopolamine
Dipyrridamole (+ oxazepam)	Persumbran	Persantin	
Doxepin	Sinequan	Sinequan	Sinequan Adapin

Generic name	Trade names		
	Federal Republic of Germany	United Kingdom	United States of America
Etilefrine	Effortil		
Fenetyline	Captagon		
Flunitrazepam	Rohypnol	Rohypnol	
Flupenthixol	Fluanxol Deanxit (+Melitracen)	Depixol Fluanxol	
Fluphenazine	Dapotum Lyogen	Modecate Moditen Motipress Motival	Permitil Prolixin
Flurazepam	Dalmadorm	Dalmane	Dalmane
Frusemide	Lasix	Aluzine Diumide Diuresal Dryptal Frumil Frusene Frusetic Frusid Lasikal Lasilactone Lasipressin Lasix Lasoride	Lasix
Haloperidol	Haldol	Dosic Fortunan Haldol Serenace	Haldol
Imipramine	Tofranil	Tofranil	Tofranil
Levodopa	L-Dopa	Brocadopa Larodopa	Larodopa

Generic name	Trade names		
	Federal Republic of Germany	United Kingdom	United States of America
Levomepro-mazine	Nozinan	Methotri-meprazine*	
Lithium salts	Quilonorm	Camcolit Liskonum Litarex Phasal Priadel	Eskalith Lithane Lithobid Lithonate Lithotabs
Lofepramine	Gamonil	Gamanil	
Lorazepam	Merlit Tavor Temesta	Almazine Ativan	Ativan
Lormetazepam	Noctamid	Noctamid*	
Maprotiline	Ludiomil	Ludiomil	Ludiomil
Melitracene	Dixeran Trausabun		
Melitracene (+ Flupenthixol)	Deanxit		
Melperone	Buronil Eunerpan		
Meprobamate	Pertranquil	Equagesic Equanil Tenavoid	Equanil
Mianserin	Tolvon	Bolvidon Norval	
Neostigmine	Prostigmin	Prostigmin	Prostigmin
Nitrazepam	Mogadon	Mogadon Nitrados Noctesed Remnos Somnite Surem Unisomnia	

Generic name	Trade names		
	Federal Republic of Germany	United Kingdom	United States of America
Oxazepam	Adumbran Anxiolit Praxiten	Oxanid	Serax
Pimozide	Orap	Orap	
Piracetam	Cerebrosteril Nootropil Normabrain		
Procaine	Novanaest	Pronestyl	Novocain
Procyclidine	Kemadrin	Arpicoline Kemadrin	Kemadrin
Pyridostygmine	Mestinon	Mestinon	Mestinon
Pyritinol	Encephabol		
Selegiline	Movergan	Eldepryl	
Strophanthin	Kombetin	Strophanthin-K *	
Sulpiride	Dogmatil	Dolmatil Sulpitil	
Thioridazine	Mellereten Melleril	Melleril	Mellaril
Tiaprid	Delpral Tiapridex		
Tranylcypromine	Parnate	Parnate Parstelin	Parnate
Tranylcypromine (+Trifluoperazine)	Jatrosom	Stelazine	
Trihexylphenidyl	Artane		
Zuclopenthixol	Cisordinol	Clopixol	

* No longer available in the U.K.

Abbreviations

A = Adrenaline
AADC = Aromatic amino acid (dopa) decarboxylase
ACh = Acetylcholine
AChE = Acetylcholinesterase
CAT = Choline acetyltransferase
CNS = Central nervous system
COMT = Catecholamine-O-methyltransferase
DA = Dopamine
DOPAC = 3,4-Dihydroxyphenylacetic acid
DOPS = 3,4-Dihydroxyphenylserine
GABA = γ-Aminobutyric acid
GAD = Glutamate decarboxylase
5-HIAA = 5-Hydroxyindole acetic acid
5-HT = 5-Hydroxytryptamine, Serotonin
5-HTP = 5-Hydroxytryptophan
HVA = Homovanillic acid
MAO = Monoamine oxidase
MHPG = 3-Methoxy-4-hydroxyphenethylene glycol
NA = Noradrenaline
PEA = β-Phenylethylamine
SDAT = Senile dementia of the Alzheimer type
TH = Tyrosine hydroxylase
Try = Tryptophan
Tyr = Tyrosine
VMA = Vanilmandelic acid

General introduction and definitions

The term "neurotransmitter" was coined by Elliot as long ago as 1904 to describe chemical compounds which are stored in the nerve cells of the brain, and indeed of other organs, and which are released from their stores by physiological or pathological stimuli. After release the neurotransmitter molecules cross the synaptic cleft, i.e. the gap between the nerve fibres, to act upon the receptors, where they produce their specific effects. This transmission of nervous activity may involve voluntary movement of some kind, involuntary actions such as breathing, digestion etc., an emotion or a change of mood, intellectual or creative thought processes or the experiencing of some kind of sensory input.

Neurotransmitter systems

There are various groups of neurotransmitter systems: 1. catecholaminergic; 2. serotonergic; 3. cholinergic; 4. amino acid dependent (e.g. GABA, glycine, glutamate, aspartate); 5. neuropeptides (modulators); and 6. histaminergic.

The catecholaminergic systems include dopamine (DA), noradrenaline (NA) and adrenaline (A), and they are essentially the transmitters of the sympathetic system and are associated with energy consumption. Serotonin (5-HT) is probably a transmitter for the parasympathetic system, and so, possibly, is histamine.

Acetylcholine (ACh) is the cerebral cortex's transmitter, and is therefore involved in voluntary movement, sensory impression, speech and thought. In the rest of the body ACh is responsible for the transmission of the nerve impulse from the motor-nerve root to the muscle end-plate for the initiation of muscular activity. ACh is a parasympathetic neurotransmitter. It is however also present in the brainstem, just as the catecholaminergic and serotonergic transmitters are not restricted to the brainstem, but also occur in the cerebral cortex. In quantitative terms, the biogenic amines NA, DA and 5-HT are however present in higher concentrations in the brainstem, because that is where the ganglia are located from which nerve tracts radiate into all the peripheral regions. There are other neurotransmitters of which we have as yet only limited knowledge, and there must be many others which we know nothing about at present, but since their clinical significance is equally poorly understood, they will only be briefly mentioned.

The catecholamines

The biosynthesis of the catecholamines is shown in outline in Fig. 1. They all originate from the essential amino acid phenylalanine. This is converted to tyrosine via the action of the enzyme phenylalanine hydroxylase, and tyrosine is converted to dopa via tyrosine hydroxylase. This enzyme is rate-limiting for the synthesis of DA, NA and A. Dopa is decarboxylated to dopamine via dopa decarboxylase, so that this enzyme catalyses the biosynthesis of a neurotransmitter amine from a precursor amino acid. Dopamine is the major neurotransmitter of the extra-pyramidal nerve tracts, governing involuntary activity, but equally emotional drive and spontaneity. The action of the enzyme dopamine β-hydroxylase leads to the neurotransmitter NA, which governs peripheral blood-pressure by its action on the heart, particularly the heart-rate. At the same time, however, it also inhibits digestive activity such as the secretion of saliva, gastric juices, bile, pancreatic juice and mucus, and peristalsis, as well as urinary output. Similarly in the lungs it leads to dilatation of the bronchi as well as inhibiting the secretion of mucus. In the central nervous system (CNS) NA is synthesised predominantly in the locus

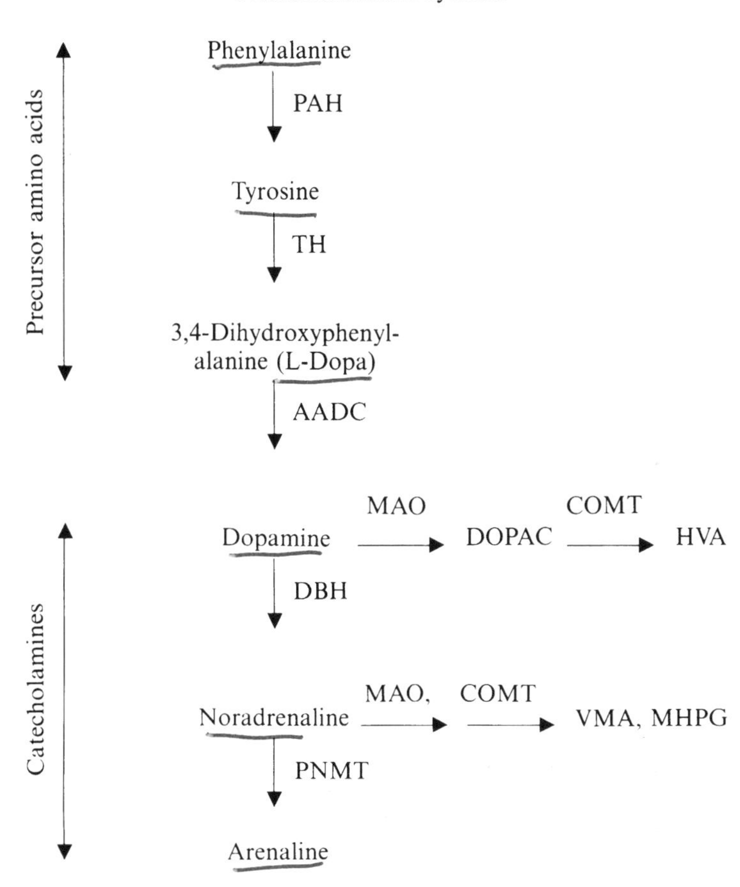

Fig. 1. Simplified scheme of catecholamine metabolism

AADC Aromatic amino acid (Dopa) decarboxylase
COMT Catecholamine-O-methyltransferase
DBH Dopamine-β-hydroxylase
DOPAC 3,4-dihydroxyphenylacetic acid
HVA Homovanillic acid
MAO Monoamine oxidase
MHPG 3-Methoxy-4-hydroxyphenethylene glycol
PNMT Phenylethanolamine-N-methyltransferase
PAH Phenylalanine hydroxylase
TH Tyrosine hydroxylase
VMA Vanilmandelic acid

caeruleus, from where it is directed, via noradrenergic pathways in the brainstem, to the limbic system, the nucleus amygdalae, the hypothalamus, the thalamus, the cerebral cortex and to many other regions of the brain. This regulates the so-called "arousal reaction", a release of NA in the mid-brain which produces an increase in cortical activity through a heightening of consciousness. Stimulation of the limbic system leads to rising emotional tension and eventually to a state of anxiety, accompanied via hypothalamic stimulation by acceleration of the heart-rate and an increase in blood-pressure. Ultimately there is an increase in peripheral muscle tone, produced by nerve impulses going down the reticulo-spinal pathways to the anterior horn cells.

Stimulation of this so-called loop results in spinal arousal and leads to increased muscle tone, setting the organism up for the biological emergency reaction ('fight or flight reflex'). As part of this reflex the release of NA in the hypothalamus leads to a lowering of body temperature, and in the thalamus to a lowering of the pain threshold.

Serotonin (5-HT)

Fig. 2 gives an outline of the biosynthetic pathway leading to serotonin, which is a sleeping-releasing substance in the mid-brain. Inhibition of the reticular formation leads to a lowering of consciousness, and any soothing effects of psychotherapy seem to act through 5-HT. This applies equally to auto-suggestive sleep training, hypnosis or sleep therapy. Furthermore, 5-HT raises the pain-threshold: this is presumably because 5-HT inhibits "arousal" functions. In the intestinal tract 5-HT induces secretory activity as well as peristalsis. Deficiency in 5-HT leads to constipation and in extreme cases to ileal blockage. In the urinary tract too, 5-HT governs urine flow. At birth the process of delivery is activated by 5-HT, whereas NA blocks it (birth anxiety). In the lungs 5-HT stimulates secretion of mucus and constriction of the bronchi, and so 5-HT is what

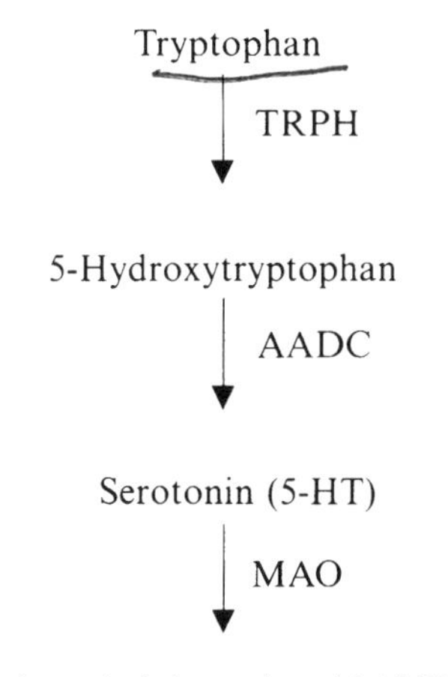

Fig. 2. Simplified scheme of serotonin metabolism

TRPH Tryptophan hydroxylase
AADC Aromatic amino acid decarboxylase
MAO Monoamine oxidase

precipitates an asthmatic attack. Finally, 5-HT lowers blood pressure. Pregnancy is a time when there is a tendency for 5-HT activity to predominate, with an increased sleep requirement, lassitude, weight gain, oedema, and a certain degree of lowering of mental activity, all of which can be referred back to a shifting of the balance in favour of 5-HT. Stress-producing factors in the environment lead to production of NA which has adverse effects during pregnancy, a time when tranquility should reign.

Acetylcholine (ACh)

Acetylcholine is the one transmitter substance that is distributed throughout the whole of the organism, and its biosynthetic pathway is shown in Fig. 3. Stores of ACh occur particularly

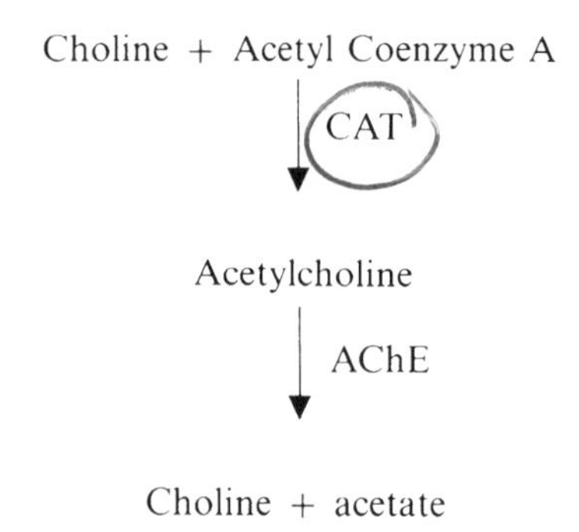

Fig. 3. Simplified scheme of acetylcholine metabolism

CAT Choline acetyltransferase
AChE Acetylcholinesterase

in the cerebral cortex, but some are also found in the brainstem. ACh is above all a neurotransmitter of fast reactions. For example if there is a relative preponderance of acetylcholinesterase (Fig. 3, AChE) in the periphery, leading to a deficiency of ACh, then the disorder myasthenia gravis will develop. This is a condition where AChE at the motor end-plates hydrolyses ACh so rapidly that premature insufficiency of muscular contraction occurs. Physostygmine inhibits the enzyme so that muscular activity is prolonged and improved. Recently animal experiments, as well as human investigations on old people, have demonstrated a decline in the activity of the synthetic enzyme choline acetyltransferase (CAT in Figure 3) in all cortical regions but particularly in the temporal lobe. This deficiency is responsible for corresponding deficiencies in higher cerebral functions such as the critical faculties, clear insight, good judgement. The Klüver-Bucy phenomenon has shown that removal of the temporal lobe leads to a total loss of memory. This indicates that the temporal lobe – and ACh – play a decisive role in the recall of memory traces.

Histamine

Histamine was the first biogenic amine to be discovered, but its functions in the CNS have probably received the least investigation. We do know, from animal experiments, that in stressed animals there is an increase in the histamine concentration of the hypothalamus, but it is not clear whether this is a specific stress-response or merely a reaction to an increased NA release. In the periphery, when insect bites cause an allergic skin reaction locally, histamine is involved in the pain, swelling and reddening, and these symptoms are lessened by treatment with antihistamine drugs. Histamine may be associated with 5-HT because it stimulates the secretion of gastric juices.

In the brain the so-called "histamine headache" is considered to be caused by histamine release, and this effect, as in the periphery, may be accompanied by a 5-HT stimulating mechanism, 5-HT also elicits oedema, and cerebral oedema certainly leads to headache.

GABA

Gamma aminobutyric acid (GABA, see Fig. 4) can also be regarded as a neurotransmitter. We recognise a GABA-ergic nerve pathway leading from the striatum to the substantia nigra which acts in an inhibitory fashion on the release of DA from the substantia nigra to the striatum. This feedback mechanism regulates any DA excess in the striatum and by GABA-ergic stimulation of the substantia nigra leads to inhibition of the nigro-striatal pathway.

Substance P

The neuropeptide, substance P, is found in the CNS and also peripherally, and we ascribe a certain stimulatory effect on the dopaminergic system to it. In Parkinson's disease there is a decrease in the substance P concentration of the substantia

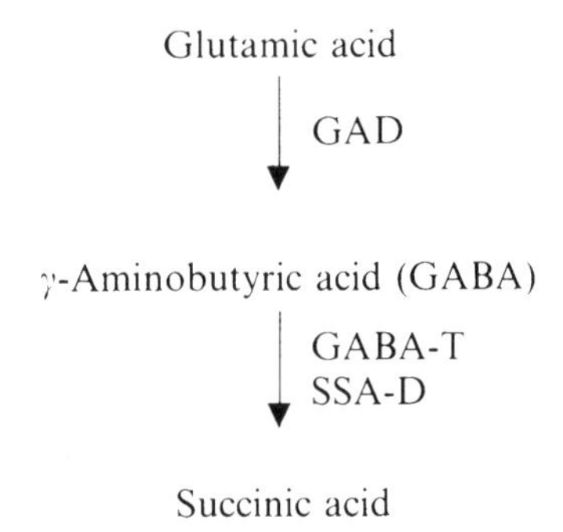

Fig. 4. Simplified scheme of GABA metabolism

GAD Glutamic acid decarboxylase
GABA γ-Aminobutyric acid
GABA-T GABA Transaminase
SSA-D Succinic acid semialdehyde dehydrogenase

nigra, but the precise significance of this is not clear. The difficulty lies in the fact that neuropeptides do not readily cross membranes, i.e. that they cannot get through the so-called "blood-brain barrier". With other neurotransmitters we know which of their precursors can cross this barrier (L-Dopa for DA, NA & A; tryptophan for 5-HT), but for neuropetides the precursors which pass the barrier have not been established, and this area of investigation still offers extensive research opportunities. However, we can say that within the organism, whenever we have found increased levels of some biochemical compound, we have eventually been able to ascribe some physiological function to that compound; examples abound, such as insulin in the pancreas, the sex hormones in the gonads, and DA in basal ganglia, all of which regulate certain definite functions.

Regulatory mechanisms

The various neurotransmitters occur in various regions of the CNS within definite concentration ranges. They are synthesised from their precursors in the appropriate nerve cells, and when they have done their job they are broken down again. The intraneuronal range of physiological concentrations depends on the balance between the synthetic and degradative enzymes, and a rise or fall in the activity of the individual enzymes thus leads to a change in function. Thus inhibition of monoamine oxidase (MAO) leads to a rise of DA in the nerve cell and consequently to a increase in those functions specific to DA. There is therefore a dynamic equilibrium between synthesis and degradation of the transmitter substance. There is mutual regulation between different neurotransmitters in order to maintain homeostasis: for example the release of too much 5-HT may lead to an increase in the activity of the antagonistically acting NA, and any pathological effects that may have arisen only disappear again on the re-establishment of the correct balance between these two transmitters.

We call such a displacement of the equilibrium a "biochemical decompensation", and its re-establishment a recompensation, and the way to such a recompensation is through existing intra- and interneuronal regulatory mechanisms. However, where the neurones have suffered degenerative changes, such feedback regulation is made more difficult if not impossible, depending on the degree of permanent impairment. As an example, where the dopaminergic neurones in the substantia nigra have degenerated, only small amounts of DA are synthesised and passed to the specific neurones in the striatum, and the consequence is akinesia, a reduced motor ability.

A biochemical decompensation produces a compensatory feedback regulation: thus a deficiency of DA in a nerve cell may be reduced by an intraneural compensatory increase in the DA-producing activity of tyrosine hydroxylase. If this is insufficient, then additional correction to re-establish the equilibrium may be made by inter-neuronal transmitter feedback coupling.

Let us look at a practical example: there are two different kinds of sleep, deep sleep and REM ("rapid eye-movement") sleep, the latter occurring during episodes of dreaming, and these two sleep phases alternate. Deep sleep is predominantly initiated by 5-HT; however there is the risk that sleep may become too deep. In deep sleep the co-ordinated lowering of blood-pressure, of heart-rate, and of blood sugar level, and the relaxation of the muscles, can all lead to decompensation, and this is the stimulus for the starting of a feedback mechanism, very likely an outpouring of NA. Now dreams begin, and blood pressure, heart-rate, breathing and muscular tone all increase. This compensatory phase, also called "paradoxical" sleep, is initiated by release of NA, which re-establishes the balance and prevents any descent into a pathological, too profound sleep situation. In old age such REM sleep (paradoxical sleep, dream states) becomes more frequent, leading to more frequent interruptions of deep sleep and thus preventing any danger of cerebral malnutrition.

We must be careful not to fall into the error of taking too simplistic a view of the balance between the neurotransmitters, such as that of Eppinger and Hess (1910) who viewed the equilibrium between the sympathetic nervous system and the vagus in terms of the beam of a balance. A more realistic kind of model might be the solar system, where the various planets, Mercury, Venus, Earth, Mars, Jupiter, Saturn etc. each move in their various individual orbits, but maintain a mutual state of tension in which attraction and repulsion may vary somewhat but strictly within closely defined values. The cybernetic rules governing such a system are in some ways analogous to those governing the maintenance of neuronal systems. If equilibrium cannot be reached within the nerve cell, then superior interneuronal mechanisms of compensation may begin to operate, and through a release of other neurotransmitters re-establish the correct level of the neurotransmitter regulatory cycle. The essential feature of our biochemical model is that the various neurotransmitters are always available to be thrown into the breach in order to maintain or to restore the balance through a negative feedback mechanism. The biochemical events within the nerve cell are a micro-model of the overall macro-cosmic events within the different brain regions. What are the possibilities for this overall balance of the neurotransmitters to be disturbed, and for this loss of balance to lead to pathological states of function and behaviour?

The neuron

If we consider, say, a dopaminergic neuron, this has a certain biosynthetic capability, and so an increase in the activity of tyrosine hydroxylase (TH) may lead to increased production and storage of DA; conversely, deficient TH activity results in insufficient storage of DA. The clinical effects of the latter effect may lead, in Parkinson's disease, to akinesia; in endogenous depression to a reduction in drive, and in old people to diminished motor and emotional activity.

The degradative enzyme for biogenic amines in neurons is monoamine oxidase, MAO, of which two types, MAO-A and MAO-B are known. MAO-A breaks down NA; MAO-B breaks down 5-HT, DA and phenylethylamine (PEA). In old age there is a relative increase in MAO-B, and this leads to a decreased storage of the corresponding neurotransmitters in the nerve cells. The clinical consequences of this approximate essentially to the negative effects of old age: loss of sleep, loss of emotional activity, reduced motor activity. The self-steering for the maintenance of the biochemical equilibria via feedback regulatory mechanisms only works as long as the appropriate structural elements are available. To take one example, in a normal healthy person it is not possible to produce hyperkinesia or psychotic behaviour by the administration of excessive amounts of the neurotransmitter precursor L-Dopa. In a patient with Parkinson's disease on the other hand, high doses of L-Dopa *can* lead to such psychotic disturbances (confusion, hallucinations, deranged thinking).

How does this happen?

When dopaminergic nerve cells progressively degenerate, for whatever reason, then there is progressively less scope for dopamine, synthesised from Dopa, to be stored. If then the availability of DA is substantially increased, then it may displace 5-HT and NA from their specific neurones. This scenario may be significant as a pathogenic component of the so-called pharmaco-toxic psychoses. Administration of tryptophan, the antagonist of Dopa and the precursor of 5-HT, then successfully competes with L-Dopa at the blood-brain-barrier active-transport sites. Further, DA stored in inappropriate neurones is displaced by the increasingly available 5-HT, so that the psychosis is lifted; however, the akinesia may persist.

This example shows that a shift in the equilibrium may generate pathological symptoms, and that these disappear again when the equilibrium is restored.

Our difficulty in interpreting the processes that are occurring within our micro-solar system are due to the fact that when we try to determine what factors may be responsible for a disturbance of the balance within the cell, we can identify and biochemically analyse only a small number of the many neurotransmitters which may be involved. There must be a host of substances which can alter the biochemical equilibrium besides those we have already met, for example we know very little about the large group of the neuropeptides. Some of them are known to be reduced in concentration in the brain-stem of patients with Parkinson's disease, without our being able to make any definite statements about the clinical significance of this finding. The neuropeptides are probably acting as modulators of the biogenic neurotransmitters.

The receptors too play a very important role in keeping the balance. As fas as the clinician is concerned it is enough to appreciate that there are post-synaptic receptors which briefly bind the neurotransmitters released from the nerve endings. In situations such as Parkinson's disease, where there is a functional lack of DA, it was found possible to stimulate the post-synaptic receptors directly by administration of apomorphine, but its adverse side-effects (particularly vomiting and kidney damage) prevented its use in treatment. Various newer drugs have been tried out over more than a decade. Apart from bromocriptine and lisuride no other drug has found widespread application. It is assumed that where there is an insufficient release of the neurotransmitter, the post-synaptic receptors become more sensitive (hypersensitivity). This makes sense, because the required function now has to be achieved with a diminished amount of transmitter substance. But negative feedback results in any stimulation leading to inhibition of TH activity, and this in turn leads to inhibition of DA synthesis. If one uses neuroleptic drugs such as haloperidol to block the receptor, then tyrosine hydroxylase activity is increased again, but this may lead to overproduction of the neurotransmitter DA. The well known effects of such overproduction include hyperkinesia, writhing movement cramps and even actual aggressive behaviour.

In addition to the post-synaptic receptors, there are also pre-synaptic receptors, which are located in the pre-synaptic membrane, from where they respond to stimuli by closing down TH activity, and thus reducing the rate of DA and NA production. One of the objectives of treatment therefore is still to try to find ways of stimulating the pre-synaptic and post-synaptic receptors independently of one another. The side effects of our attempts to achieve stimulation of the post-synaptic receptors remain much as they were in the first phase of the introduction of treatment with L-Dopa (1960–1965). The most serious of them is orthostatic hypotension: the systolic blood pressure on standing up from a sitting or recumbent position may fall to as low as 50 mm Hg. In spite of these drawbacks, there can be no doubt of the value of postsynaptic receptor stimulation for the re-establishing of the biochemical balance in dopaminergic systems. Despite the relatively few therapeutic strategies available to the clinician, it is nowadays very often possible to design corrective strategies for a patient through biochemical decompensation.

What is so fascinating in our biochemical balance model is that when we come to consider the origin of the various symptoms, our model effectively eliminates the distinction that has hitherto been made between somatic and psychic symptoms. A preponderance of NA can lead on the one hand to the somatic symptoms of rapid heart-rate, increased blood-pressure, dry mouth, cessation of intestinal peristalsis etc. and on the other to the psychic symptoms of excitement, agitation, anxiety and insomnia. In other words the chemical neurotransmitter is able to activate not only somatic but also psychic functions. The *chemical transmitter is thus the link between the age-old antithesis of soma and psyche.* This realisation is particularly vital to the treatment of psychosomatic illnesses.

Let us take as an example the case of a woman with a neurosis involving compulsive washing of the hands. This compulsion was for her a release from a permanent state of anxiety and insomnia. She underwent three years of psychoanalysis, after which she came to me (W.B.) and told me in despair: "I now realise that what has led to my present state of anxiety and insomnia was the fact that as a child my mother punished me for playing with my clitoris, and made me feel guilty. But this realisation is no help at all!" Our interpretation would be along the lines that someone who has had a severe traumatic experience in childhood has produced a cybernetically distorted pattern within the brain. This acts like an interfering radio transmitter which "jams" subsequent mental development and inhibits the process of maturation through a distortion of the balance of the neurotransmitters, resulting in a psychic decompensation. The genius of Sigmund Freud lay in the fact that without having available any of our stock of knowledge about neurotransmitters, he was yet able to demonstrate that somatic as well as psychic symptoms may be caused by traumatic states of frustration, that these symptoms are not under our conscious cortical control, and that we therefore cannot free ourselves from them by conscious intellectual effort. In fact he recognised that the brain-stem was the seat of defective regulation. If he were alive today, the classical psychoanalytical method as practised nowadays would probably bring a smile to his face . . .

We must, from the outset, keep our attitude to the relationship between body and mind firmly in view. There is no doubt that psychological stress, whether within the family or at the workplace, if it is not coped with in some suitable fashion, can lead to ulcerative colitis, bronchial asthma, high blood pressure etc. Continuing stress first of all produces danger-signs such as insomnia, gastric problems or headache. After all, stress is nothing more than the body's attempt to come to terms with the biochemical decompensation caused by an adverse environment. At first this attempt is made calmly; but when the possibilities of achieveing a balance from one's own resources are exhausted, then it is likely that some symptom denoting a disturbance of the equilibrium between the transmitters will appear. The bad conscience of a marriage partner after an infidelity will only lead to a loss of appetite, to insomnia or to depression, if the biochemical resources for an adequate biological compensation of the guilt feelings are lacking. This does not obviate the need for psychotherapy either in general or in specific cases. When, for example, a sick patient is enabled to sleep again, it is immaterial whether this is achieved by way of acupuncture, or after a sympathetic chat, after an instructive interview or after administration of tryptophan (precursor of 5-HT): success is invariably achieved through a restoration of the biochemical equilibrium. Thus the calming effect of a reassuring interview, of auto-suggestion, of psychoanalysis, may release serotonergic effects which contribute to the disappearance of specific symptoms.

Of course psychological treatment cannot cope with profound structural damage, it can only be a supportive measure:

a severely depressed patient also needs anti-depressive drugs to assist with psychic guidance. Psychotherapy by itself cannot do it, biochemical compensation cannot always achieve it either, but in our modern society it is often more successful because it is more rapid in action. The harassed executive who is overweight, who has high blood pressure and a threatened coronary, who has dizzy spells, who gets tired easily and has trouble sleeping, is hardly likely to get straightened out by a course of psychoanalysis, but a carefully designed drug regime may chemically diminish his raised NA-activity.

An example of long-term adjustment of the biochemical balance is environmental lighting. The intensity of the light source exerts an action on the (dopaminergic and) noradrenergic neurons of the midbrain through the dopaminergic system of the retina. This produces an arousal reaction, i.e. the sleeper wakes up, and the cerebral cortex is activated and becomes responsive to emotional and motor stimuli. In winter the environmental lighting is lower, but in the spring it brightens again, and induces impulses which activate the DA- and NA-producing systems. The brighter lighting intensifies motor activity, mental alertness and sexual stimulation. In the autumn the lowering of the lighting level has the opposite effects, and promotes greater readiness to sleep, to feelings of lethargy, and to mental lassitude, which if it goes to pathological extremes may result in endogenous depression.

The ability to adjust to changes in the weather is likewise governed by neurotransmitters. Patients with lesions of the brainstem, whether traumatic (e.g. caused by traffic accidents), inflammatory (e.g. encephalitis), or vascular (deficient CNS blood-flow), all suffer from a diminished ability to adjust to meteorological changes. The well known phenomenon of the "Föhn-sickness" (hypersensitivity to the warm alpine Föhn-wind of the spring) may have various effects on its victims: some react with a hyperactivity of the catecholaminergic transmitters (aggressive behaviour, increased proneness to road accidents, tendency to pick fights); others which react with a hyperactivity of inhibitory transmitters such as 5-HT, involving greater sleep requirements, inactivity, lethargy and apathy. People react differently, according to their inherent neuronal state.

In old age there is usually a reduction in the concentrations of the neurotransmitters in the neurones, compared with the concentrations occuring during youth. The availability of DA as well as of NA and 5-HT is lower in old age, and this may well be the cause of the diminished adaptability that occurs with advancing years.

Biological decompensations caused by individual neurotransmitters

Noradrenaline (The neurotransmitter of the whole sympathetic nervous system) is responsible in the CNS for the ability to be aroused, i.e. for vigilance.

Hyperactivity: in the periphery leads to rapid heart-rate, high blood pressure and muscular cramps; insomnia, weight loss (pubertal anorexia), irritability, agitated behaviour, restlessness and anxiety. The pain-threshold for all types of pain is lowered.

Reduced activity: low blood pressure, slowed pulse, slack posture, lack of initiative and decisiveness, increased tendency of fatigue and apathy, as after severe exhaustion following an infective illness.

Serotonin (Possibly a neurotransmitter of the parasympathetic nervous system) in the CNS is sleep-inducing, emotionally calming; lowers blood pressure, increases heart-rate.

Hyperactivity: improved appetite, weight increase, urge to sleep, consciousness rather damped down, basically unenterprising mood, diarrhoea or constipation. Slowed thought processes, loss of drive, decreased muscle tone, slowing of the circulation, oedema or tendency to thrombosis, increased pain threshold.

Reduced activity: sleeping badly, poor posture, introverted, no urge to activity.

Dopamine The neurotransmitter for all the extrapyramidal motor tracts of the CNS (for all instinctual movements) for muscle tone (static & dynamic) inhibition of all trophic functions.

Hyperactivity: choreic movements, compulsive nervous movements ("fidgety Phil"), emotional hyperactivity, tonic cramps – especially at night, hyperactive limbic system, tendency to anorexia.

Reduced activity: hypokinesia or akinesia, hunched attitude, tendency to physical weariness.

This brief outline rather resembles a mediaeval map of the earth, full of blank spaces. Nervertheless, there are well-explored areas of biochemical and clinical certainty concerning normal and pathological function, as well as the biochemically established factual knowledge about the synthetic and degradative enzymes of the neurotransmitters inside nerve cells. This knowledge is sufficient to give us access to specific ways of achieving replacement therapy. This may be by administration of neurotransmitter precursors which can get through the blood-brain-barrier, by inhibition of the specific degradative enzymes, or by directly stimulating the various specific receptors.

In any therapeutic strategy these are the three approaches to be considered for re-stablishing the equilibrium, and whose application in a multipolar therapy must aim at a mutual balancing of the neurotransmitters.

A simple analogy might be a "mobile", one of those structures made from wires connected with fine threads. At the end of each thread some object is attached (in a Christmas mobile these objects might be angels, stars, shiny balls, etc.). These objects all move about but are carefully balanced against one another. But if one of the objects drops off, then the whole mobile slews round or hangs crooked.

For each brain region we could construct such a "mobile" of neurotransmitters, which must normally float around in a state of balance. However, this balance is affected by influences from other brain regions and their "mobiles". This last mechanism is particularly significant, because it can compensate for a disturbance of the "mobile" in another part of the brain. Because this compensation can be spread over several adjacent brain regions and their systems, none of these is normally ever overloaded.

Sites of action of psychoactive drugs and treatment strategies

The class of drugs described as psychoactive ought really to be described as "neuroactive", because their effect on the psyche is achieved through the nervous system.

Figures 1–2 are diagrammatic representations of the synthesis and degradation of the most important neurotransmitters, compounds correctly described as "biogenic amines", which are synthesised from their precursors within the neurones (ganglion cells). They are broken down inside the cell by MAO; and outside the cell to some extent by MAO, but mainly by COMT (serotonin is an exception to this), and eventually excreted. The precursors are amino acids such as tyrosine (Tyr) or tryptophan (Try), which can easily cross the blood-brain-barrier. A schematic representation of the processes that take place within a noradrenergic ganglion cell is shown in Fig. 5. Tyrosine (Tyr) is brought to the cell in the circulation

---

Fig. 5. Schematic representation of chemical transmission in a noradrenergic synapse. Noradrenaline is produced from tyrosine in a three-stage enzymatic synthesis, and stored in synaptic vesicles. The nerve impulse produces an influx of calcium ions and these induce the release of noradrenaline from the vesicles into the synaptic cleft. The released transmitter binds to a specific receptor-protein embedded in the postsynaptic membrane and initiates a series of reactions in the receiving neurone, for example short-term effects via electrical impulses, and also long-term effects. The action of noradrenaline is terminated by a series of mechanisms including rapid re-uptake into the nerve ending, and enzymatic breakdown. The release of noradrenaline into the synaptic cleft also activates presynaptic receptors at the end of the axon, and this initiates the production of cyclic AMP which activates protein kinase and thus regulates noradrenaline biosynthesis. [From Iversen LL (1979) The chemistry of the brain. In: The brain. Freeman, San Francisco, pp 70–81]

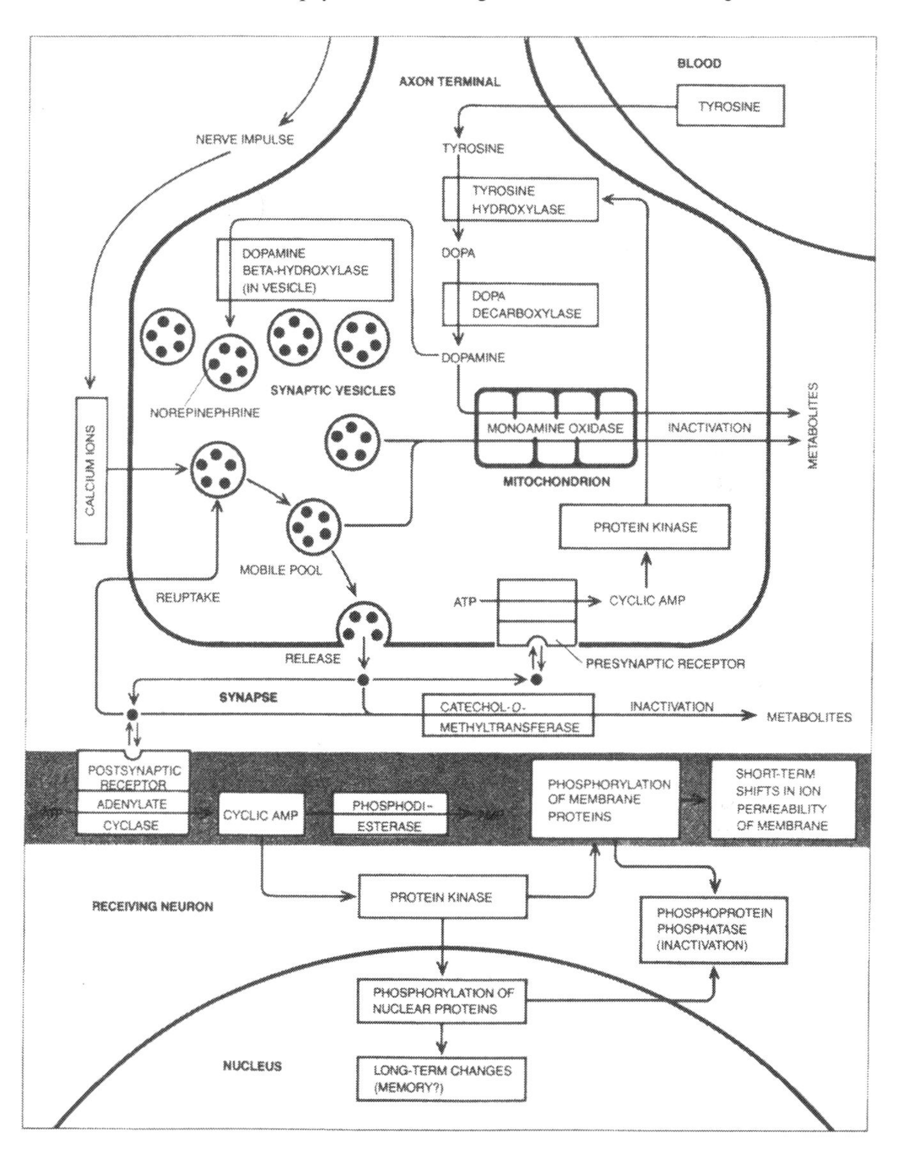
AXON TERMINAL
BLOOD
TYROSINE
NERVE IMPULSE
TYROSINE
TYROSINE HYDROXYLASE
DOPAMINE BETA-HYDROXYLASE (IN VESICLE)
DOPA
DOPA DECARBOXYLASE
DOPAMINE
SYNAPTIC VESICLES
NOREPINEPHRINE
CALCIUM IONS
MONOAMINE OXIDASE
INACTIVATION
METABOLITES
MITOCHONDRION
PROTEIN KINASE
MOBILE POOL
REUPTAKE
ATP
CYCLIC AMP
RELEASE
PRESYNAPTIC RECEPTOR
SYNAPSE
CATECHOL-O-METHYLTRANSFERASE
INACTIVATION
METABOLITES
POSTSYNAPTIC RECEPTOR
ADENYLATE CYCLASE
CYCLIC AMP
PHOSPHODI-ESTERASE
PHOSPHORYLATION OF MEMBRANE PROTEINS
SHORT-TERM SHIFTS IN ION PERMEABILITY OF MEMBRANE
RECEIVING NEURON
PROTEIN KINASE
PHOSPHOPROTEIN PHOSPHATASE (INACTIVATION)
PHOSPHORYLATION OF NUCLEAR PROTEINS
NUCLEUS
LONG-TERM CHANGES (MEMORY?)

and enters through the cell membrane. In the cell tyrosine hydroxylase (TH) catalyses its conversion to Dopa, which is decarboxylated to dopamine by aromatic amino acid decarboxylase (AADC). Noradrenaline is synthesised from dopamine in specific ganglion cells through the activity of dopamine-β-hydroxylase (DBH). In a similar fashion, tryptophan (Try) is converted to 5-hydroxytryptophan (5-HTP) in serotonergic ganglion cells, and AADC decarboxylates 5-HTP to serotonin (5-HT). In the brain-stem these biogenic amines, DA, NA and 5-HT, are stored particularly in selectively localised cell bodies. DA is stored in dopaminergic neurones, is released by physiological or pharmacological stimuli, crosses the synaptic cleft and arrives at the post-synaptic receptors, where it exerts its physiological activity (muscular action, raising muscle tone). Similarly, NA is the neurotransmitter which in the brain-stem is responsible for arousal: i.e. wakefulness and a heightening of consciousness. 5-HT is the transmitter which in the brain-stem induces sleep and promotes repose, and in peripheral regions governs the whole process of digestion from salivation through secretion of digestive enzymes to intestinal peristalsis.

These classical neurotransmitters, and indeed others both known and as yet hardly suspected, regulate all involuntary processes, all the moods and emotional reactions and all extrapyramidal motor functions. They are thus the deciding factors for the way we feel and the way we act. The synthesis and degradation of these neurotransmitters is regulated via negative feedback mechanisms. Too much DA in a neuron acts as a brake on the activity of TH, an action mediated via autoreceptors. Conversely, a depressed DA level elicits an increase in TH-activity and leads to an increased DA synthesis and an increased neuronal DA content. However, feedback mechanisms also work to regulate mutual correlations between the individual neurotransmitters. These mutual relationships, between NA and 5-HT, between DA and 5-HT and between DA and NA therefore are also functionally in equilibrium. A higher or lower level of one neurotransmitter can be neutralised by a coupled feedback process and brought back in physiological equilibrium. The multidimensional dynamic equilibrium of all the neurotransmitters governs our capacity for achievement and our ability to recover. Environmental stimuli may upset this balance, either in the short or in the long term, which may manifest itself by the appearance of clinical symptoms. Upsetting events during the day liberate increased amounts of NA, and so disturb the balance: too much NA and too little 5-HT tend to result in insomnia, loss of appetite, and a dry mouth. On the other hand, too much rich food and alcohol can release increased amounts of 5-HT, with the symptoms of a flushed face, perspiration, tiredness and somnolence, together with a lack of drive and resolution. Too great an outflow of DA makes a person twitchy and fidgety, but too little outflow of DA leads to lassitude and a slouched posture.

The psychoactive drugs intervene in this interplay of the neurotransmitters. It is therefore important to the practising physician to understand their sites of action and their designed effects, because a carefully planned and rational mode of treatment will be more successful for the patient and more satisfying to his doctor.

Principal groups

1. Tranquilizers
2. Stimulants
3. Antidepressant drugs
4. Neuroleptic drugs
5. MAO-inhibitors

Tranquilizers

The tranquilizers are the most widely prescribed drugs, because they provide rapid relief. The benzodiazepines (e.g. diazepam, nitrazepam, lorazepam etc.) are the best known.

Tranquilizers bind to specific receptors and so exert an action on their normal neurotransmitters. Benzodiazepine receptors are linked to GABA-ergic neurotransmission, and in this way exert a modulating effect on the actions of the "classical" neurotransmitters. To give an example, anxiety releases NA, but if the excessive release of NA can be inhibited, then the patient no longer consciously experiences the anxiety. Both patient and doctor would prefer the disturbing symptom e.g. anxiety, insomnia, or whatever it may be, to disappear. Since tranquilizers directly and indirectly modulate many neurotransmitter systems, they not only liberate the patient from the target-symptom "anxiety", but unfortunately they also lower blood pressure, make the patient slump, dampen down awareness of reality and diminish attention-span and the ability to concentrate. These side-effects may in some circumstances overshadow the desired effect to such a degree that the treatment loses its effectiveness. The enormous world-wide usage of tranquilizers is a sign that our modern life-style, with its emphasis on continuous achievement, demands a correspondingly increased NA-turnover. The tendency of our times is against the idea that we should contain our desire for achievement to where we can calmly deal with the resulting chemical imbalance so that individuals can reach an equilibrium about the particular degree of stress that lies within their own physiological ability to handle. The tendency is rather for people to depend on a drug to reestablish their feeling of well-being, very possibly at a neurotransmitter level that may not be physiologically appropriate, and may therefore lead them to a long-term use of tranquilizers. Since our behaviour is not always governed by common-sense, but may depend on unconscious instinctual behaviour originating in the brain-stem, it is only in *acute* situations (insomnia, anxiety) that a tranquilizer is actually indicated, because then it can immediately restore the equilibrium.

Psychiatrists in institutions often talk of tranquilizers such as lorazepam causing dependence. In our view this is not true.

There are patients with psychopathic constitutions who cannot quietly and on their own restore their neurotransmitter imbalance, and such people respond to every problem by reaching for some kind of tranquilizer, whether it be alcohol or lorazepam. However, true dependence has two main features: continually increasing dose requirement to achieve the same effect, and the occurrence of withdrawal symptoms when coming off the drug. Neither of these apply to the tranquilizers. Admittedly people get used to them: elderly patients who regularly take a certain dose of a particular tranquilizer get so used to it, that they need it to get to sleep. The same applies to the glass of wine that an old gentleman may need as a night-cap. In order to avoid patients getting used to tranquilizers in this way it is essential that they are only prescribed to settle an acute problem, such as sleeplessness just before an exam, or excessive grief following a bereavement or fear of flying. In endogenous depression, the states of anxiety can initially be suppressed with a tranquilizer, and this has the advantage of rapid action. But one has to be aware that a tranquilizer cannot cure the depression, and that it is therefore necessary also to prescribe an antidepressant drug at the same time, although it may take longer for it to attain the desired effect.

Stimulants

These are drugs which can release neurotransmitters from ganglion cells and therefore potentiate their actions. Best known are amines that produce wakefulness, such as the amphetamines. These deplete the nerve cells of their neurotransmitters and at the same time block reabsorption. This instantly leads to a substantial increase in the activity of a neurotransmitter, because so much more remains available to exert its normal action. However, the depletion of neurotransmitter stores results in the potentiated action being followed by a phase of exhaustion and depression. The fatal end of the racing cyclists who dope themselves with amphetamine stimulants is well known and makes it clear that particularly in situations demanding prolonged performance, such stimulants are contra-indicated. On the other hand we need not be too hard on the actor or the politician who pops the occasional pill to sharpen up his performance. However, in the first place it should not become a habit, and in the second place one must not delude oneself that it can overcome exhaustion: a stimulant can only tap available energy. It is just the same as with coffee, one of the earliest stimulants: the effect of the caffeine it contains only improves performance when the organism is rested. If the subject is already overtired, then all that happens is that coffee puts him in a bad mood or makes him jittery, without improving his performance. In practice stimulants are not widely prescribed because they only have a transient effect, and because their euphoric effect never seems to make up for the subsequent depressive phase. Even among the young of the "drug culture" there is relatively little abuse of rapid-acting drugs like "speed" and "bennies".

Antidepressant Drugs

The first effective antidepressant, imipramine, was discovered by R. Kuhn in Switzerland. However, its mood lightening effect takes some weeks to make itself felt. The mode of action of antidepressant drugs is to block re-uptake of the neurotransmitters into their stores in the nerve cell, and thus to increase their concentration within the synaptic cleft, which down-regulates postsynaptic receptors and improves mood and drive. Some antidepressants are more effective in increasing initiative, others are more effective in relieving anxiety. We were able to establish that clinically speaking all antidepressant drugs work through their ability to restore a balance. A good antidepressive improves drive via its effect in potentiating the actions of dopamine, it improves resolve and decisiveness through potentiation of the activity of NA, and at the same time it improves the patient's sleep via the potentiation of 5-HT.

Antidepressant drugs only have relatively minor side-effects, and do not cause addiction. After the problems have gone, the patients generally stop taking the drugs off their own bat, sometimes prematurely. Occasionally patients complain about dryness of the mouth, but it is usually tolerable. More serious, perhaps particularly in women, is the tendency to put on weight, which is viewed with disfavour precisely during the remission period. Patients tend to put on weight especially with those antidepressants which produce an increase in 5-HT, and this promotes an improved appetite and an associated tendency to put on weight.

Neuroleptic Drugs

The most effective and best known neuroleptic is haloperidol. Neuroleptics work primarily by blocking receptors. The result is a state of absolute rest, which in a tense aggressive schizophrenic is exactly what is wanted, and neuroleptics are capable of rapidly damping down even a raging alcoholic. Overall, the practising physician has relatively little call to use a neuroleptic. It may have side-effects such as depressive disturbances and may contribute to the appearance of negative extrapyramidal symptoms such as a Parkinson-like akinesia or positive symptoms such as torsion-dystonic cramps. Blocking DA-receptors is what elicits the Parkinson-like symptoms, which may appear as retarded and delayed ability to move, even after only a few days on the drug. The DA-receptor blockade produces stimulation of TH, with the consequent increased rate of DA-synthesis, via feedback regulation, and this upsets the balance between dopaminergic and cholinergic function and may lead to positive extrapyramidal symptoms, e.g. the "tongue and throat" syndrome, where there is a compulsive twisting of the neck and sticking out of the tongue. Another possible result is the "restless legs" syndrome, where the patient finds it impossible to keep the legs still, whether sitting down or standing up. Such side-effects are readily dealt with by the additional administration of anti-cholinergic drugs. The practising physician does not often need to prescribe neuroleptic drugs, but problems may arise where an individual neuroleptic drug is promoted by a drug company on the basis that it may help to reduce the dosage of tranquilizers. Thus fluphenazine, thioridazine and pimozide are in fact neuroleptics, and their long-term administration can eventually produce extrapyramidal symptoms. Schizophrenic patients who, when they were in institutions, got accustomed to particular neuroleptic regimes based on drugs like fluphenazine, zuclopenthixol, flupenthixol etc., can be given the same drug intramuscularly by their own doctor once or twice a month without any problems except that an anticholinergic drug (biperidene or something similar) must be given at the same time. One can recommend doctors to put one or two ampoules of haloperidol into their "black bags" in case of seizures of all kinds, as well as one or two ampoules of diazepam (10 mg), which is very effective in less severe overexcitement or anxiety.

Melperon is particularly useful for elderly patients who become confused or agitated at night, because it is followed by less of a hangover the next morning.

Powerful neuroleptic drugs do achieve the desired sedative effect, but tend to be followed the next morning by reduced blood-flow in the brain and a resulting lethargy and apathy. While we may recommend practising physicians to use tranquilizers and antidepressive drugs on their patients, we also caution them to be very sparing in their use of neuroleptics.

The so-called "tardive" dyskinesias that occur in older patients who have been on neuroleptic drugs for several years are evidence that such long-term medication may produce not only biochemical but also permanent structural changes. Treatment of such long-term side-effects is not at all satisfactory, which argues that they are the result of a deep-seated lesion in neuronal metabolism.

MAO-Inhibitors

Monoamine oxidase (MAO) is an enzyme which contributes to regulation of the dynamic equilibrium, both intra- and extraneuronally, by degradation of the biogenic amine neurotransmitters, whether they be DA, NA or 5-HT. We have had drugs that inhibit MAO for over twenty years, but the early drugs were associated with too many side-effect, so that today they are of only historical interest.

In 1968 Johnston discovered that there were different forms of MAO, and this led to the development of specific MAO-inhibitors. Tranylcypromine, a non-specific inhibitor, predominantly blocks the breakdown of NA and 5-HT. Selegiline, a specific inhibitor of MAO-B, predominantly blocks degradation of DA and phenylethylamine. While MAO-A inhibitors at higher doses produce a rise in blood-pressure, which may be particularly threatening after ingestion of cheese and other foodstuffs because of the potentiation of the action of the pressor amine tyramine which many cheeses contain, the use of selegiline completely avoids this "cheese effect". It is therefore the drug of choice for achieving a tonic effect in all DA-deficiency syndromes such as Parkinson's disease, depression, pubertal anorexia nervosa, and generally in old age.

Tranylcypromine is used in low doses as a stimulant in retarded depression, and helps in childhood depressions, difficulties with concentration and attention span, mood disturbances and premature fatigue. An ideal preparation consists of tranylcypromine in combination with trifluoperazine (a neuroleptic drug). In the package-inserts supplied with MAO-inhibitors there is always a statement to the effect that a combination with tricyclic antidepressives is not permissible. This no longer holds when sensible dose-levels are used and the patient is informed about the need to be careful about diet. W.B. has been using a combination of parstelin (one tablet in the morning) with amitriptyline in the evening as the ideal antidepressive prescription in medium grade depressions. Tranylcypromine in contrast to selegiline, raises the blood-pressure.

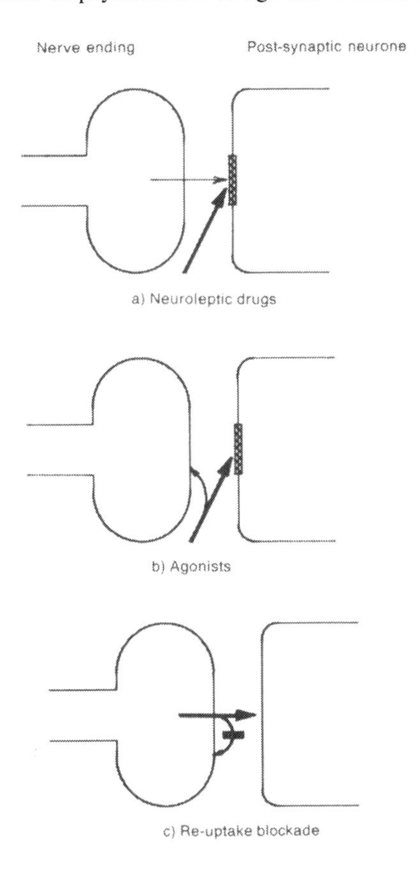

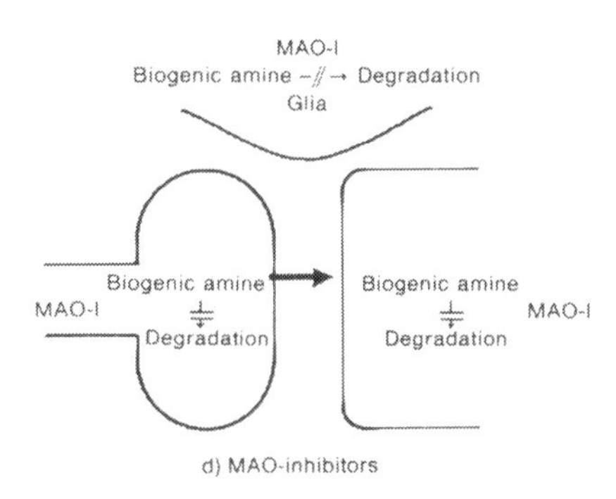

d) MAO-inhibitors

This brief overview of the modes of action of some psychoactive drugs (see also the diagrammatic representations in Fig. 6), is meant to assist the practising physician in his daily professional activities to select the correct measures for reestablishing the patients' biochemical equilibria. The observations and experience in treating patients should enable him or her to make the right choice of the optimal medication for the symptoms of a particular patient without reference to package inserts. Since on the average 40–50% of the working day will in fact be concerned with patients with some sort of psychological problem, this knowledge and experience could mean a fruitful increase in the proportion of successful treatments.

Fig. 6. Simplified representation of the modes of action of important drugs

a Neuroleptic drugs block post-synaptic (dopaminergic) receptors, so that the neurotransmitter (dopamine) released at the nerve ending is unable to exert its activity. (Main application: e.g. in the treatment of psychoses)

b "Agonists" predominantly stimulate post-synaptic receptors directly, and thus improve or replace the physiological stimulus normally elicited by the natural messenger substance. (Main application: e.g. in the treatment of Parkinson's disease)

c Re-uptake-blockers inhibit the re-uptake of neurotransmitter released into the synaptic cleft (biogenic amines, for example) and thus potentiate the action of the natural messenger. (Main application: e.g. depression)

d Monoamine oxidase inhibitors block the intra- or extra-neuronal breakdown of biogenic amines and thus potentiate the physiological effects of the neurotransmitters. (Main application: e.g. in depression or in Parkinson's disease)

Pain

The most comprehensive information-delivery system, and one which has the most unpleasant effects on our consciousness, concerns our sense impressions of pain. We distinguish two types of pain sensations: *epicritic,* meaning that the pain is sharp, distinct and localised, and *protopathic,* meaning that it is dull, not precisely definable, diffuse and deep. Specific pain chemoreceptors react to a variety of substances: plasma kinins, bradykinin, histamine, serotonin, prostaglandins and neuroleptic drugs, while non-specific mechanisms may amplify our subjective impressions of pain. Our individual constitutions, together with the circumstantial effects of the different neurotransmitters, can significantly alter the way we actually feel pain: NA increases our perception of pain, and the NA-associated anxiety likewise contributes to our apprehension of pain. Heightening of consciousness (the "arousal reaction") also increases sensitivity to pain. Ischemia raises the sensitivity of the pain receptors in the affected part. Thus there is considerable variability in the pain threshold. People with a sympathotonic constitution tend to feel pain more, while those of a "vagal" type appear to have their pain threshold set higher. Circumstances leading to a stimulation of NA, lower the pain threshold, and these circumstances include fever as in infections, and also hyperthyreoses. The influence of serotonin (5-HT) with its sleep-association tends to raise the pain threshold.

The clear "epicritic" type of pain is conducted via A-delta myelinated nerve-fibres, while the dull "protopathic" pain impression is conducted by C-fibres. These C-fibres are afferent and run alongside the sympathetic nerve fibres, reaching the spinal cord without switching through any ganglia. In the hind-brain the first junction is in the substantia gelatinosa. Substance P has been found in high concentrations in the hind-brain, and plays a part in the transmission of pain impression. A variety of other neuropeptides are involved in the transmission of pain: somatostatin and the endorphins appear to inhibit pain sensation, while neurotensin promotes it. In the spino-thalamic tract A-delta nerve fibres conduct the sharp pain sensations. In the evolutionarily more ancient region of the spino-thalamic tract C-fibres conduct the protopathic, dull, diffuse, harder to localise feelings of pain.

It is quite understandable that an essential sensation like pain can be modulated at various stages. Its function is to notify our consciousness of the existence of danger, and is thus an "arousal" function. In the brainstem all sensory pathways, particularly the pain fibres, send branching fibres to the reticular formation and this produces cortical activation and heightening of consciousness, but also stimulates the limbic system, which evokes both emotional reactions such as anxiety and involuntary reactions such as acceleration of the heart-rate, increased blood-pressure and, through the descending nigro-reticular pathways, a stimulation of muscular tone. We may assume that NA is the neurotransmitter of the generalized "arousal reaction", since the emotional and involuntary responses and the increased muscular tone all indicate this. Of course neuropeptides with a protective action against pain, such as neurotensin, the prostaglandins or somatostatin probably exert a modifying influence. In a similar way, pain-inhibiting factors such as endorphins or enkephalins may have a sedative effect on pain sensation. The neurotransmitter associated with such inhibitory actions might well be serotonin, and this biogenic amine raises the pain threshold. What is fundamental is that pain sensation is converted in the brainstem to a radiation of impulses to the limbic system, and elicits the corresponding affective-emotional and also involuntary reactions. At the same time in the brainstem feedback reactions go into action, which inhibit an excessive response to

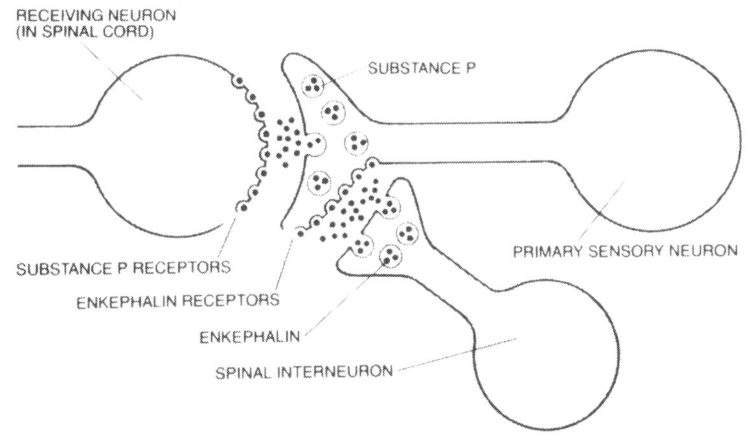

Fig. 7. Hypothetical model of the mechanism for the transmission of the nerve impulse at the first synaptic relay of the spinal cord, showing the transmission to the brain of a pain signal at a peripheral pain-receptor. The dorsal root of the spinal cord has enkephalin-containing interneurones, coupled to the axonal ends of a pain-receptor neurone which contains substance P as its transmitter. Release of enkephalin from the interneurone inhibits the release of substance P, so that the spinal receptor neurone receives a diminished excitatory stimulus and so sends fewer pain messages to the brain. Opiates such as morphine can bind to the unoccupied enkephalin-receptors and thus mimic the pain-suppressing effect of the enkephalin system. [From Iversen LL (1979) The chemistry of the brain. In: The brain. Freeman, San Francisco, pp 70–81]

painful stimuli. There is no doubt that release of serotonin into this region, which can produce numbness and even loss of consciousness, inhibits any dramatic and excessive pain sensation. Conscious experience of pain is not only dependent on the quantitative degree of the painful stimulus and on the strength of the arousal reaction triggered by the transmitter NA acting together with pain-enhancing neuropeptides, but also on the inhibiting effect on the arousal reaction of transmitters such as serotonin acting in concert with opiate agonists like enkephalin and endorphin. This model of the way that the pain process can be modulated is not just an intellectual construct, but rests on firm clinical experience.

There exists a kind of "safety cut-out" which diverts unbearable pain, capable of completely destroying any such maintenance of the internal equilibrium, into unconsciousness, so that the victim "passes out", and this neutralises life-threatening pain. As far as treatment goes, the more peripheral the place where one can neutralise the painful stimulus, the more successful the treatment. The pain of a local periostitis (tennis elbow) can be removed by simple infiltration of procaine. However, when pain penetrates into the reticular formation then its radiation into the affective and involuntary area of the limbic system is so extensive that only opiate agonists are able to reduce it. Apart from physiological neuropeptides there are only a few drugs that are capable of neutralising these radiating decompensations, and one has to have recourse to tranquilizers, antidepressive drugs or neuroleptics, depending on the severity of the pain.

Generally speaking one can say that protopathic pain sensation (muscular cramps, vascular cramps) have the ability to disturb the patient in the depths of his being. The sharply defined epicritical pain of e.g. a slipped disc has quite a different effect. Clinically speaking it is not hard to diagnose such pain. This kind of pain sensation originates in the skin, in the legs, muscles, connective tissues and joints. The deep, diffuse protopathic pain cannot be localised easily like that: it has a dull, aching character and is also not transmitted so rapidly as for example the ligthning stab of a trigeminal neuralgia. It is fluctuating both in its arrival and its departure, and it correlates with vegetative symptoms that accompany it, such as sweating, rapid heart-beat, and goose-pimples, as well as subjective sensations such as anxiety and dread. Pain which affects only one side of the body is usually of central origin (thalamus), while pain which affects both sides suggests a spinal origin, or even a polyneuropathy (diabetes). Segmental pain is the result of a mechanical cause, such as slipped discs or inflammatory reactions (neuritis). The protopathic forms of deep pain are produced by cramps, by stretching of the hollow organs such as stomach, intestines, gallbladder, urinary bladder, urethra etc. Vascular pains too belong to this group, and they are mediated at the pain receptors by histamine, serotonin and the prostaglandins.

An interesting manifestation of a protopathic kind of pain is causalgia, which consists of a burning sensation in one of the extremities. It was particularly apparent in the last war as a result of gunshot wounds of the extremities. Characteristic features of causalgia are burning pains, and the very dry, red and thin skin of the affected extremity. The so-called "summation effect" is also very characteristic: the slightest touch triggers a paroxysm of pain of volcanic proportions; even the banging of a window or a shout can act as a non-specific trigger. During the war the treatment of choice was sympathectomy of the appropriate segment. Nowadays one does not see such kinds of pain so often, and one also commands the whole range of modern psychopharmaceutic drugs from bromazepam to haloperidol.

What one does come across more and more often nowadays is the "burning feet" syndrome. It consists of burning pains in the lower extremities occurring most often at night in the warm bed. The bedclothes are pulled off the legs and the feet are constantly on the move. One may assume that this is caused by a parasympathetic activation of some kind in the limbic system: at least the flushing and the warming up of the legs strongly suggest this. It does not appear that the burning feet are actually the site of origin of the pain: rather they may be the site to which a limbic dysfunction refers the pain. In a mild case treatment consists only of walking about, and the stimulation this affords seems to be enough to restore the biochemical balance. More serious cases may require bromazepam or amitriptyline. The "burning feet" syndrome is not infrequently a symptom of a disguised depression, a sign of a biochemical neurotransmitter imbalance especially of the limbic system.

Such a condition may lead to projections into various other organs, for example a gall bladder cramp, or a frequent need to urinate or to chronic constipation. Such projected distortions of the balance demand a carefully chosen antidepressive drug.

A pain symptom which often turns up is brachialgia paraesthetica nocturna, a painful paraesthesia in one or both hands, mostly occuring at night. If it only affects one hand the patient's fear that it might be some kind of stroke can rapidly be dispelled, it is only necessary for him to move or massage the affected hand and the loss of sensation and the pain diappear at once, which would not be the case with a transient ischaemic attack. The immediate cause of this cervical syndrome is usually a cervical spondylopathy, causing a pathological alteration of muscular tone. These disturbances of muscle tone may be started off by muscular cramps in the neck muscles or by a slack hypotonia of the same region (typing, piano-playing). A reduction in tone may be achieved with bromazepam or baclofen, while a tonic effect on the muscles may be effected with L-dopa together with a peripheral decarboxylase inhibitor such as benserazide.

The muscular tone of the whole back and hip region is actually the key to understanding all of the many problems of back-pain. The so-called "slipped disc" is the classic model for a sharp accurately localised pain. A sudden twisting or bending action in the loin region of the spine causes an immediate segmental pain, radiating from the crutch down to the toes; post-traumatic oedema makes it worse, and this can lead to escalation in the motor roots. The patients feel a weakness in the affected leg, and the reflexes are lacking. A loss of feeling for all kinds of stimuli (touch, pain, heat) permits precise localisation of the affected segment of the spine. Surgical intervention is indicated when this loss of feeling gets worse or leads to complete anaesthesia, otherwise an acute occurrence may be treated by physical manipulation, which is often able to restore normal posture. The pain may disappear by itself, or it may require several carefully placed infiltrations of procaine, to which it may be a good idea to add cortisone in order to reduce the oedema.

Far more common than this dramatic traumatic model with its acute presentation is the more usual disc problem caused by overwork, or by excessive exertions in sport, where a relaxation of muscular tone in the loin region comes about through fatigue or weariness. There follows slowly increasing pain in the appropriate segment of the spine, which only occasionally radiates as far as the ankle. This is again a sharply localised sort of pain, which the patient can point too precisely, and which he can intensify or diminish by bending to the left or to the right or forward. The primary cause of this kind of slipped disc is an insufficiency of muscular tone, caused by a defect in the neurotransmitters responsible, which are mainly acetylcholine, dopamine and noradrenaline.

These are the reasons why so many situations involving deficiencies of transmitter substances are associated with loin pain. If one gets over-tired after prolonged walking or hiking or climbing, the muscle tone deterioration elicits a pain reaction. In viral infections such as influenza there may also be a reduction in muscle tone, which again makes latent pain syndromes flare up. In a depressive illness too, a relative lack of transmitters leads to loss of muscle tone, causing a hunched, stooping posture and the flaring up of loin pain. In the extreme case of Parkinson's disease the lack of dopamine and noradrenaline eventually produces the typical stooped attitude which leads to mechanical irritation of the lumbar-sacral nerve-roots and the typical pain associated with this irritation. Similar considerations explain the secundary pain that may occur in the region of the cervical spine. Cervical migraine is the best-known example. It comprises the pain at the base of the neck and across the shoulders that occurs in the morning and that can radiate diffusely into the crown and forehead and even to the eyes. The origins of this pain may be a pathological barrier to movement in the cervical spine which causes irritation of the vertebral arteries and a restricted blood flow in the region of the basilar artery. Knotting of the neck muscles and a slack muscular tone associated with a hidden depression may also cause a reduced blood supply to the basilar region and an ischaemia-associated headache. The key to cervical migraine and similar types of headache is the stability of the cervical spine. If it is only a case of knotting of the neck muscles, massage, with or without a muscle relaxant such as baclofen, may be enough to take away the pain. The same effect can be achieved by prescribing an antidepressive tonic (e.g. melitracen, dibenzipine or tranylcypromine plus trifluoperazine) to stimulate the poor tone of the neck muscles.

This brings us to modern man's most common problem, the *headache,* which must rank as the one pain that affects everyone at some time or another. Countless books have been written about headache, and scores of hypotheses proposed to explain it, but objectively confirmable ways of classifying the possible or probable causes are lacking. Of course there are some headaches which can be readily explained, such as those due to intracranial pressure (tumour, cerebral oedema associated with encephalitits or a post-traumatic episode such as concussion). Increased intracranial pressure stimulates the pain receptors, and the dispersion of the extra water with human albumin, cortison or diuretics gives temporary or permanent relief.

Another type of headache which can be fairly well classified is classical *migraine.* This typically one-sided headache with nausea and vomiting generally starts at night or in the small hours of the morning, and may last for hours or even days. As one of the triggers a change in the weather is often responsible, usually the passage of a front followed by an area of low pressure. This can be interpreted by postulating that low atmospheric pressure results in release of serotonin. In normal people this may only cause slight fatigue, or sleepiness or a mild loss of drive, but people prone to migraine are less able to cope with this release of serotonin, and so we can visualize the occurrence of a migraine because of the dilatation of the arterio-venous anastomoses in the brain, and this lets the oxygenated arterial blood flow away through the veins leaving the brain. We may then regard the hypoxia as the immediate cause of the migraine pain. In the so-called "*migraine accompagnée*" there are also associated speech disturbances, and one-sided muscular weakness, loss of sensation and vision, and also scintillating scotoma.

From a clinical point of view, changes in the weather as triggering factors for a migraine attack are plausible explanations, and represent the migraine attack as a disturbance of certain neurotransmitters. In fact, according to my own experience (WB), long-term medication with noradrenaline-stimulating antidepressive drugs such as melitracen gives better results in the long run than the usual migraine treatments, whose aim is merely to cope with the symptoms of a disturbance, which is always harder than the prophylactic treatment aimed at preventing a neurotransmitter distortion from occurring in the first place.

A less common form of headache is the *cluster-headache,* which occurs mainly at night, is one-sided, and leads to flushing of the face on the affected side. Other, less frequently occurring symptoms are bloodshot eyes, watering of the eyes, swelling up of the nasal mucosa and contraction of the pupils. As the name suggests, the symptoms are not continuous, but in a typical "cluster" may come and go several times before disappearing. We can assume that serotonin (or other neurotransmitters with a similar mode of action) are involved in this kind of headache, because all the symptoms suggest it. As rational treatment would be administration of L-tryptophan and serotonin-stimulating antidepressive drugs such as amitriptyline can be recommended.

The most frequent type of headache is the so-called *vasomotor attack*. This is an "omnium gatherum" sort of syndrome which fits into no sensible classification or analysis except that such patients seem to react to all forms of stress by having one of "their" headaches. Wind, cold, heat, infections, fatigue, alcohol, frustration, anxiety etc. are typical causes. Such a manifold collection of triggers is clear evidence of a constitutional tendency on the part of the patient. One may consider that a particular sensitivity of the pain receptors could be the common pathogenic feature. In our opinion it is not possible to prescribe a generally effective treatment that covers all these diverse kinds of headache.

A fairly common form of headache is the *tension-headache,* often characterized by the sufferers as the "band of steel round the head" which feels like a painful constriction. The source of this is taken to be a painful knotting-up of the neck musculature. Tranquilizers, which are not only emotional but also muscle relaxants can be recommended for dealing with this kind of headache.

A particular kind of pain syndrome is *trigeminal neuralgia,* which is characterised by sudden attacks of extremely severe pain in one of the three branches of the trigeminal nerve. The pain lasts only a fraction of a second and can be produced ligthning-fast by stimulating certain definite trigger points such as touching a tooth cavity with the tip of the tongue or a morsel of food. Other triggers are chewing-, lip- or swallowing-movements and even speech. The ligthning speed, the precise localisation and the sharp severity of the pain all point to it being epicritic in nature, and attacks tend to be associated with advancing years. Because of the way this pain appears it may be defined as epileptic in character, and in fact the use of the antiepileptic drug carbamazepine is sometimes effective. Additionally we can reduce the frequency of the attacks with serotonin-stimulating antidepressive drugs (amitriptyline, doxepin and mianserin).

A special pain syndrome, which these days seems to be found in older people, is the *pain of shingles.* The specific herpes zoster virus, which is identical with the chicken-pox virus, attacks the sensitive spinal ganglia, but not all that infrequently it also attacks the motor neurons. The peripheral paralyses generally disappear over a period of months. The segmentally localised areas affected by the virus are the sites of very acute pain. The spinal pain-receptors are stimulated by the inflammation to such an extent that, almost spontaneously, acute pain sensations radiate from the shingles scars towards the centre without any apparent causative stimulus. This pain is almost unbearable, especially at night. It seems directly paradoxical that during the serotonin-phase of night, in which the pain threshold is inherently higher, the shingles pain persists as a continuous and unbearable irritation. It indicates, as mentioned earlier, that other neurotransmitters are responsible for pain. Treatment is limited to tranquilizers, and in particularly dramatic cases, the use of haloperidol. A critical fact about the pain associated with shingles is that attempts at modern rapid treatment are often not pressed home sufficiently strongly. At the first sign of a segmental involvement, with severe pain, one must immediately use infusions of amantadine, together with additional cortisone. At the same time, supplementation with gamma-globulin can be helpful. Nowadays the former treatment of shingles, with vitamin-B injections and powder-bandages must be condemned as totally inadequate.

Sleep

The biological phenomenon of "sleep" is an instinctual action, an active process, which represents an energy-restoring function for the organism, and which is entirely governed by the mid-brain. W. R. Hess showed in animal experiments that cats whose mid-brain region was electrically stimulated fell asleep, and Jouvet succeeded in isolating the putative neurotransmitter responsible for releasing sleep: it turned out to be serotonin (5-HT). He was also able to show that para-chlorophenylalanine, an antagonist of 5-HT-synthesis, could stop an animal from falling asleep. The precursors tryptophan and 5-hydroxytryptophan, both of which can cross the blood-brain-barrier (see Fig. 2), are converted to serotonin in the nerve cell. Release of the neurotransmitter then induces sleep. The switching off of the conscious waking state goes hand in hand with a total biological transformation of the whole organism: the phase of sympathetic performance which is maintained by the transmitter noradrenaline (NA) is replaced by the energy-restoring recovery phase: blood-pressure is lowered, the pulse is slowed, breathing becomes shallower and the pain-threshold is raised.

Recently supra-optic centres (the suprachiasmatic nucleus) have come to be recognised as the centres which govern sleep. These are the same centres which von Economo in the twenties held responsible for governing sleep. Destruction of these centres leads to loss of the sleep-rhythm. On plane journeys, where one is rapidly crossing time-zones, the sleep phase adapts to the dark-light rhythms, but the endogenous rhythm persists for a time, and this leads to the feeling of disorientation called jet-lag. Flying west accentuates REM-sleep (see below for a definition of REM-sleep), while flying east blocks REM-sleep, and

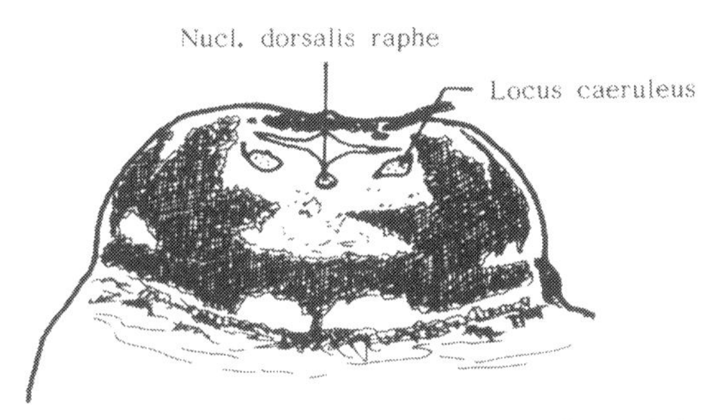

Fig. 8. Brain regions important for the regulation of the sleeping-waking cycle. The locus caeruleus contains noradrenaline-synthesising cells, those of the raphe nucleus synthesise serotonin

this has a stronger disorienting effect and leads to feelings of fatigue, lethargy and indecision. As well as the normal 24-hour diurnal rhythm there is also a short-term 90-minute fluctuation, which runs day and night independently of the light-dark cycle, and this is regulated through centres in the pons.

A non-REM phase of sleep is governed by the dorsal raphe nucleus in the same region, and this phase is responsible for the restorative functions of sleep. It is followed after about 90 minutes by a REM-activity phase (i.e. dreaming) which is activated by the locus caeruleus. This leads to an energy-promoting phase. These two nuclei maintain the 90-minute cycle via feedback mechanisms between building up of energy (raphe nucleus) and energy consumption (locus caeruleus), and this 90-minute cycle is quite independent of the 24-hour cycle governed by the supra-optic centres.

Presumably the 90-minute cycle is the archetypical and more ancient rhythm, and it is apparent in our daily life only in a rudimentary form. Only arousal is limited to 90 minutes, and a renewed phase of activity, attributable to the release of adrenaline, only becomes possible after an interval for the restoration of endogenous energy stores. All genuine sleep states commence with a phase of non-REM sleep.

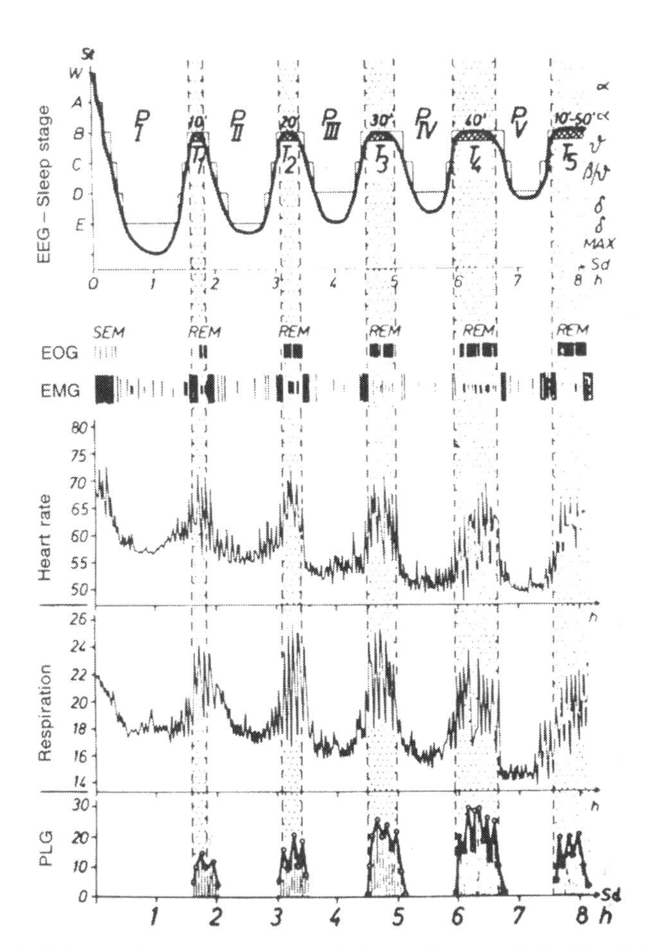

Fig. 9. Schematic representation of the various periods ($P_I - P_V$) of one man's night's sleep. From the onset of sleep to waking in the morning the dream periods ($T_I - T_V$) get progressively longer, and so do the intervening periods of deep sleep. Heart-rate and respiration slow down progressively up to the seventh hour, but show periodic increases during the dream episodes.

SEM slow eye movements
REM rapid eye movements
PLG phallography (occurrence, duration and variations of penile erections)
h Duration of sleep in hours

[From Jovanovic UJ (1976) Sleep and the autonomic nervous system. In: Sturm A, Birkmayer W (eds) Klinische Pathologie des vegetativen Nervensystems, vol 1. G Fischer, Stuttgart, pp 363–450]

We can divide phases of sleep into *deep sleep* phases, which are elicited *inter alia* by the release of e.g. serotonin, and the interspersed *REM (rapid eye movement) phases,* which among other stimuli, are elicited by e.g. noradrenaline. REM-sleep is dream-sleep, which coincides with activation of the EEG, of heart rate, of breathing and of blood pressure (Fig. 9). A REM phase – also called paradoxical sleep because it is not elicited by serotonin, but rather by NA – is a feedback mechanism, i.e. when 5-HT-sleep gets too deep, a counteracting force comes into play leading to NA-activation which then restores the biochemical 5-HT/NA balance, and this highly significant auto-regulation guarantees a recuperative sleep equilibrium. A permanent state of deep-sleep would eventually lead to a coma, in which arousal to a state of consciousness would become too difficult. Dreams are therefore the expression of a cerebral activation caused *inter alia* by NA, which however rarely proceeds as far as waking up into full consciousness. With advancing age these REM-sleep phases become more frequent and this is a useful feature which prevents deviation into a threatening energy imbalance. Any lesions of the locus caeruleus, the source of NA-production, leads to a reduction in the proportion of REM-sleep and thus to an insufficiency of all cerebral activities.

The balance between energy consumption and energy production

The interneuronal equilibrium between e.g. the sleep-producing transmitter 5-HT and the arousal-transmitter NA, which is maintained by negative feedback mechanisms (parasympathetic/sympathetic equilibrium) is the prerequisite for the correct adjustment of the balance between energy consumption and energy restoration. When this balance is lost, then one experiences disturbances of the sleep-wake-rhythm. These disturbances may result from disturbances of the internal milieu as well as by disturbances caused by the external environment.

Endogenous disturbances were described in detail by von Economo during the course of his investigations on patients with the *"sleepy sickness"* (*encephalitis lethargica*) after the first world-war. He observed patients who were unable to sleep, and others who slept for weeks on end. Nowadays we also have to consider patients with sleep disorders caused by various other types of encephalitis or by brain tumours.

Sleep-disturbing factors and possibilities of treatment

1. Environmental stimuli

Basically any stimulus from the environment may lead to a disturbance of the NA/5-HT balance. Increased illumination has effects on the retina which are still not clearly understood, but which involve melatonin, dopamine and other neurotransmitters, and lead to activation of the brain-stem where they produce an "arousal reaction". This waking reaction (Moruzzi – Magoun, 1949) is predominantly elicited by the neurotransmitter NA.

All afferent sensory stimuli originating in the periphery (sight, sound, touch, heat and pain) deliver stimuli to the NA-neurones in the reticular formation, and this stimulation produces the arousal reaction. Consciousness becomes heightened (anxiety, joy), the cycle of involuntary functions becomes activated through activation of sympathetic functions by NA (heart rate, blood pressure, drying up of mucosa), and muscular tone is increased by stimulation of the peripheral spinal gamma-loops. These are the prerequisites for our ability to overcome the most varied challenges in our lives. Just as this biological phenomenon of "arousal" is essential for us to overcome any danger, so its excessive manifestation can have adverse effects on our normal waking-sleeping-rhythm. Thus the stimulus of illumination when watching too many films or too much television may lead to sleep disturbances through a pathological arousal reaction, just as sound stimuli caused by traffic or industrial noise pollution is a common cause of sleep disturbances.

2. Peripheral painful stimuli

There is a host of painful stimuli, which can send a continuous stream of pain messages to produce an arousal reaction in the brain-stem which will adversely affect normal sleep function. They include ischialgia, spinal root neuralgia caused by a protruding intervertebral disc, the pain of shingles, the most various itches, pain in the joints caused by senile arthroses but also the ischaemic pain of occluded arteries, intestinal griping in infection and indigestion, and all kinds of cervical pain of spinal origin (especially at night).

3. Emotional stimuli

The realm of the emotions represents a wide-ranging source of disturbed sleep. Consciously experienced emotional irritation such as anger at something that has happened at work or at home, happy occasions amongst friends or family, consciously experienced prolonged tension (flying, extended car journeys) all may cause sleep disturbances.

Sleep disturbances caused by emotional or mood-related factors respond well to psychotherapeutic relaxation methods such as autosuggestion, which can effect a switch from sympathetic to parasympathetic activity. One expression of such a process is the generation of a feeling of warmth or of heaviness produced by auto-suggestion.

4. Sub-conscious stimuli

Stimuli originating in the subconscious are a comprehensive constellation of causes for sleep disorders. Guilt complexes of the infantile phase are associated with anxiety reactions originating in the whole range of neurotic functions. They all lead to an excessive sympathetic reaction whose consequence is insomnia.

5. Frustration

Frustration in one's work or on the sexual side is also capable of generating an arousal reaction via hyperactivity of sympathetic neurotransmission, with consequent insomnia. People can also be *deprived of sleep,* either through stresses in the course of their social life, or deliberately by others as in the course of "brain-washing" of political prisoners or because whole communities have been traumatised by political terrorism.

6. Consumption of stimulants

Coffee is undoubtedly the most widely used agent for holding sleep at bay; caffeine, amongst its other properties leads to raised catecholaminergic activity, and it is this that promotes wakefulness. This ability to make us feel livelier is of the greatest value in our achievement-oriented life-style, because it helps us to overcome feelings of lethargy after meals, after difficult long or boring conferences and after long car journeys. One ought to avoid taking too much coffee in the late afternoon, because it blocks the recovery phase of sleep. In severe exhaustion following vigorous physical activity, the neurones become depleted of dopamine and noradrenaline. Drinking coffee then cannot activate any transmitter functions, and this results in sleeplessness and ill-temper.

Alcohol is a non-specific releaser of neurotransmitters, and also more specifically, of 5-HT, and so it leads to a flushed face and sweating. Release of 5-HT produces tiredness, lethargy and sleep. There is nothing wrong with moderate enjoyment of alcohol, but during the daytime, particularly in the morning, alcohol has a bad effect on one's work capacity, and must be avoided in occupations where alertness must be maintained, such as driving, skiing, flying and operating machinery.

7. Depression

Of the endogenous causes of insomnia, depressions are highly significant, whereas schizophrenia-like syndromes are seldom contributory. Depressive syndromes are characterised by a combination of symptoms that include deficiency of 5-HT in the brain-stem, and this readily explains insomnia, loss of appetite and consequently of weight. The cause of the depressive ill-temper can be referred to a loss of the equilibrium between NA, DA and 5-HT. The fall in the 5-HT level in the reticular formation produces a preponderance of NA, and so we wake up too early, and lie awake with unpleasant thoughts going round and round in our heads. If, on the other hand, the level of 5-HT exceeds that of NA and DA, then we experience a lack of arousal instead; we cannot concentrate, we are out of sorts, nothing will satisfy us, and we suffer from a general ineffectuality; sleep and appetite, however, are not affected.

Many depressed patients are tormented by anxiety, and this may depend on an increased release of NA. In such a situation NA-potentiating antidepressive drugs such as melitracen, dibenzipine and tranylcypromine plus trifluoperazine make things worse by accentuating the anxiety and aggravating the insomnia. Such patients, who are suffering from agitated depression, need 5-HT-stimulating antidepressive drugs such as amitriptyline, with or without chlordiazepoxide.

8. Manic states

The most severe degree of insomnia is found in manic states. Manic patients drift around all night, make innumerable plans and start innumerable projects, none of which ever come to fruition. This sleeplessness can eventually culminate in complete exhaustion. Simple tranquilizers or sedative antidepressants are completely inadequate in such a situation: what we have to do is to prescribe neuroleptic drugs, and the ones that are indicated range from chlorprothixen via methotrimeprazine all the way to haloperidol. The last of these is the most powerfully acting neuroleptic and in cases of manic insomnia can be given via injection. It is, however, also the most effective drug in states of psychotic excitement within the schizophrenic repertoire such as the delirious agitation of alcoholics or after head injuries. The exact drug regime depends very much on the severity of the manic agitation, perhaps starting with methotrimeprazine (25–100 mg), going on to chlorprothixen (50 mg) or melperon (25–50 mg) and proceeding to flupenthixol (2–4 mg), clopenthixol (10–25 mg) or fluphenazine (5 mg) and eventually to haloperidol (1–2 mg). For patients with actual manic disease one may need to go immediately to lithium treatment (lithium carbonate, 2–4 tablets a day).

Of course neuroleptic drugs are not necessarily free of side-effects: after several days one may notice that the patient's movements tend to overshoot (hyperkinesias, tonic cramps), but a single ampoule of biperidene injected intravenously or intramuscularly is very quickly effective in overcoming such side-effects. Neuroleptic drugs frequently also cause circulatory disturbances, particularly orthostatic hypotension. The way to handle the less severe forms of deliriant confusion is with infusions of L-tryptophan (1000–3000 mg daily) or of 5-hydroxytryptophan (50 mg intravenously, but not commercially available). Meprobamate also has a sedative effect which lies between those of tryptophan and injections of neuroleptic drugs.

9. *Reduced cerebral blood supply*

Older patients suffer from a particular kind of sleeplessness, which is due to a reduced cerebral blood-supply. A compromised cerebral blood supply may give rise – especially at night – to cerebral ischaemia, which is likely to be accompanied by loss of sleep, confusion and at times even aggressive feelings and agitation. In patients such as these the sleep disturbance is not caused by excessive release of a neurotransmitter, but by insufficient nutrients reaching the brain, and so drugs which function by damping down neurotransmitter activity are not indicated. Because there is not enough oxygen getting to the brain, there is also inadequate nutrition, and this is best corrected in the first place by stimulating the circulation, and for this one can, for example, give digitalis in the evening.

Another drug which promotes sleep – especially in the elderly – is strophantin (¼ mg intravenously), and dihydroergotamine works in a similar way. In addition these older patients can also receive tranquilizers: these are the most effective and least harmful drugs in the less severe sleep disorders. Admittedly, they may lead to dependence, but not as is sometimes alleged to addiction. Most tranquilizers are benzodiazepine derivatives, including diazepam, lorazepam, clobazam, oxazepam, lormetazepam etc., and exert a sleep-promoting action.

These drugs can give very satisfactory results in older patients with sleep disorders, regardless of how the disturbance originated. They are, however, also used at lower dosages as a daytime sedative, a practice that cannot be recommended. Unrest and agitation during the daytime are often caused by a masked depression, and a sedative antidepressant drug is therefore more to the point. Tranquilizers are not only the most harmless but also the safest drugs, and it is almost impossible to commit suicide with them.

There is a particular group of tranquilizers which offer the preferred effect in promoting a healthy sleep, and these are, in order of their effectiveness, flunitrazepam (2 mg), nitrazepam (5 mg) and flurazepam (30 mg). Older people with sleep problems can take flunitrazepam for years on end, and if its effect should decline, they can be switched over to nitrazepam or flurazepam, because this seems to work better than increasing the dose of the original drug. However, if there should be a "hangover" effect the following morning (lassitude, dizziness, dullness), then the dose must be decreased.

Pathological sleep conditions

States of excessive sleeping are much rarer than conditions involving loss of sleep. As already mentioned, such excessive sleep states are associated with patients with post-encephalitic Parkinson's disease. Among my own patients (WB) I have seen several who had periodic episodes of this kind. In the so-called "Pickwick syndrome" (in children), in addition to a significant craving for fat, the patients also exhibit a sleep addiction. The logical way to treat this appeared to be stimulant amines such as amphetamine, but these proved inadequate. As one of its actions amphetamine releases NA from the neurons, and blocks its reuptake from the synaptic cleft, but this fails to re-establish the balance between NA and 5-HT. For these reasons antidepressant drugs with an activating mode of action seem to be more effective.

Climatic conditions

Climatic conditions can have the effect either of preventing sleep or of promoting it. Rainy weather, cool and still days, and conditions of low atmospheric pressure generally are more favourable to falling asleep and sleeping well and long. Dry heat, high atmospheric pressure and winds tend to inhibit sleep and lead to restlessness. There are regions noted for their health resorts which seem particularly favourable to sound sleep: the relationship between the climatic conditions there, and their effect on neurotransmitters has not been sufficiently investigated.

Sleep, as an archetypal function, may be regarded from the evolutionary viewpoint as a very ancient instinctual activity. Even very primitive life-forms still have energy-charging phases in their life-cycles. Man, the highest and most complex life-form, suffers more frequent disturbances through a multitude of overlapping feedback regulatory processes than life-forms which live less complex and relatively undisturbed lives in harmony with their circadian daynight rhythms.

It is harder to disturb the balance of archetypal functions, because they are evolutionarily ancient engrams firmly rooted in the brain-stem, but when a disturbance does occur it is also harder to restore the balance.

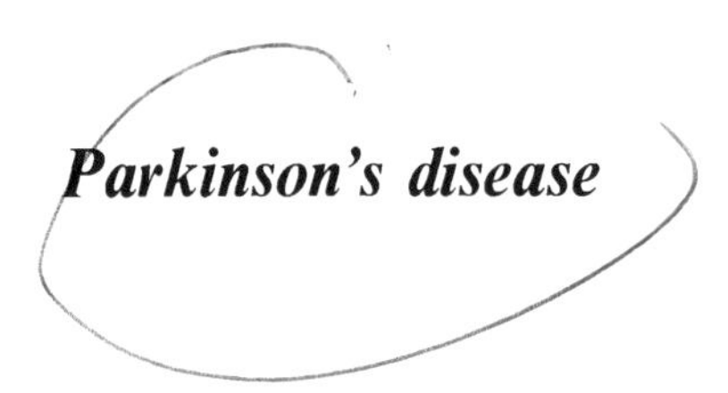

Parkinson's disease

Parkinson's disease is always described as the typical disorder that is caused by the pathological deficiency of a particular neurotransmitter. Remarkable advances have come both from biochemical analyses in the region of the brain-stem and from the therapeutic measures that were suggested by the results of those analyses.

Arvid Carlsson, starting out from the discoveries of B.B. Brodie, demonstrated that if animals were given reserpine, then the stores of dopamine in the basal ganglia were emptied; this dopamine depletion was accompanied by changes in the animals' psychomotor behaviour. By administration of L-Dopa, the precursor of dopamine, it was possible to refill the depleted stores with dopamine, and at the same time the psychomotor disturbances disappeared. This fundamental experiment was the starting point for practical research into Parkinson's disease. From a clinical viewpoint, this "dopamine deficiency model" was extended and completed in the sense that it was now also assumed that various other non-motor problems such as excessive salivation, seborrhoea, depressions, anorexia etc., were due to anomalies in other neurotransmitters. Thus, beside the DA-deficiency in the substantia nigra, the caudate nucleus, the putamen and the globus pallidus, there are also diminished NA-concentrations in the basal ganglia. Similarly, reduced levels were also determined in the hypothalamus, nucleus ruber and above all in the locus caeruleus. The serotonergic pathways are distributed similarly to the noradrenergic ones. Here too there is a deficiency, though to a different degree.

In the brain GABA is an inhibitory neurotransmitter. The appropriate neurones are in the striatum and project into the

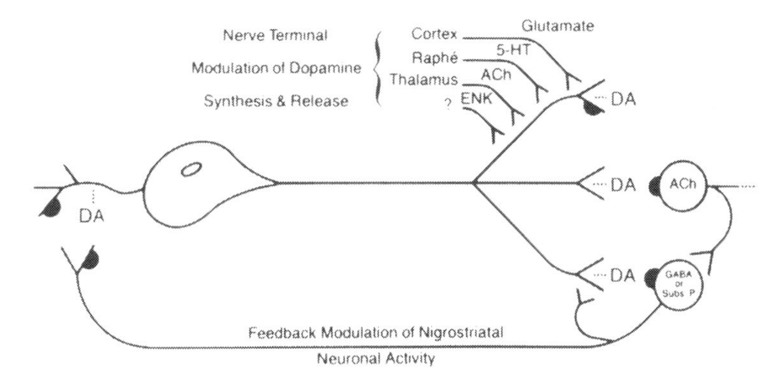

Fig. 10. Simplified representation of the nigro-striatal system of the brain. The pre-synaptic dopaminergic function can be modulated by a variety of different mechanisms. [From: Bhatnagar RK, et al. (1982) In: Dopamine receptors and their behavioural correlates. Pharmacol Biochem Behav 17 [Suppl] 1: 11–19]

substantia nigra, where GABA exerts its inhibitory effects on DA-release. Substance P can be shown to occur in increased amounts in the basal roots of the spinal cord, where it is a sign of the central pain-transmitting nerve tracts, and it is also present in increased amounts in the brain-stem, particularly the substantia nigra, and there it represents an excitatory neurotransmitter modulating the action of DA.

Neuropeptides also occur in the CNS, and their functions are at present not completely established. In animal experiments endorphin produces a rigor-like rigidity, reversible on administration of naloxone and apomorphine. 5-HT accentuates endorphin-akinesia. Endorphin also raises the 5-HT level in the raphe nucleus.

The dynamic equilibrium between the neurotransmitters and neuropeptides is maintained by various mechanisms of feedback regulation. Neuropeptides are described as co-modulators of the neurotransmitters. The activity of tyrosine hydroxylase is inhibited by stimulation of auto-receptors.

Nigro-striatal DA-activity is affected by cholinergic, GABA-ergic and peptidergic stimuli, and by signals from other neurones. Parkinson's disease results when about two thirds of the dopamine in the substantia nigra has been lost; the loss of two thirds of the cholinergic cells in the basal nucleus contributes to dementia, i.e. a deficiency of about two thirds of the neurones can just about be compensated by increased activity of the remaining one third. There can be little doubt that it is the lack of DA in the substantia nigra and the striatum which is responsible for the akinesia in Parkinson's disease. However, even in healthy people one can show that in every decade there is a 10–13% decrease in DA, and this explains the changes in motor activity that occur in healthy older people. The disappearance of DA in patients with Parkinson's disease accelerates during the course of the disease, and administration of DA is unable to halt the march of the disease process. Although initially treatment with L-Dopa (madopar, sinemet) gives excellent results, as the disease progresses so the clinical effectiveness declines. As the onward march of the disease continues, so the incidence of "off"-phases (periods when all movement is blocked) becomes more frequent and above all their duration increases. Eventually these intermittent periods of movement blockage run together into akinetic crises which may last for days or even weeks. Analysis of the brains of patients who died in such akinetic crises showed only minimal amounts of DA. There is also a decreased tyrosine hydroxylase activity which parallels the loss of DA. This loss can also be demonstrated in the limbic system, and the psycho-pathological disturbances which may appear during the course of Parkinson's disease are attributable to it. While tyrosine hydroxylase in the CNS declines steadily, regardless of the administration of therapeutic drugs, the activity of MAO (monoamine oxidase) hardly alters in Parkinson's disease. The activity of this DA-degrading enzyme exhibits diurnal variation, with a maximum in the afternoon between noon and 6 p.m., and this may contribute to the occurrence of movement blockage ("off"-phases) at about this

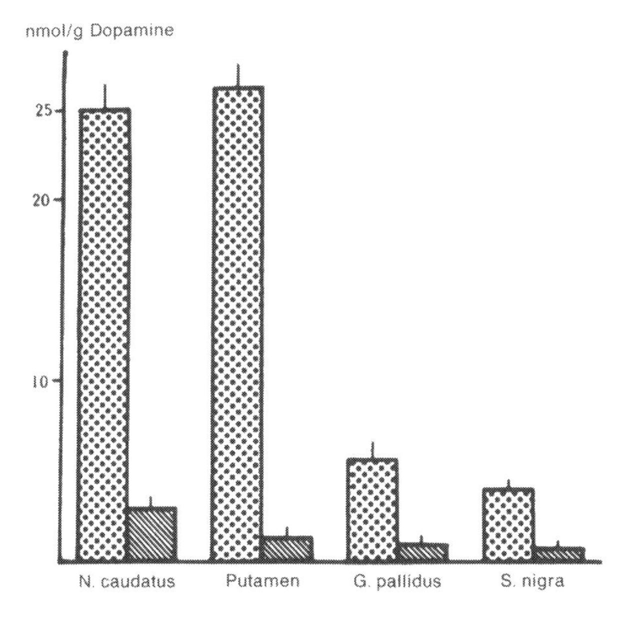

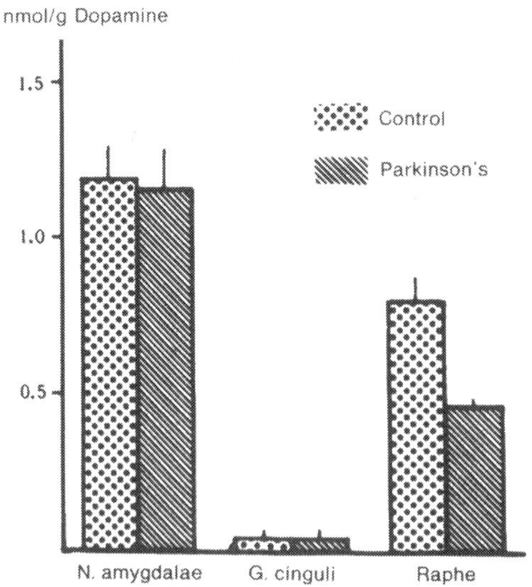

Fig. 11. The dopamine deficiencies in various regions of the brain in Parkinson's disease. [From: Birkmayer W, et al. (1977) In: Psychobiology of the striatum. Elsevier/North-Holland]

time of day. The effectiveness of selegiline, a specific MAO-inhibitor, reinforces our view that these off-phases are due to the unavailability of neuronal DA. The distribution, storage and release of DA is regulated by the pre-synaptic auto-receptors and the post-synaptic receptors via feedback mechanisms. The auto-receptors inhibit tyrosine hydroxylase, while stimulation of the post-synaptic receptors by dopaminergic substances slows down the rate of pre-synaptic firing. The activation of pre-synaptic receptors and the blocking of post-synaptic receptors therefore elicit similar types of behaviour. If there is widespread degeneration of the pre-synaptic nerve processes, then a hypersensitivity of the post-synaptic receptors may be induced. When the DA-neurones in the substantia nigra degenerate, activation of the post-synaptic receptors can compensate for the decrease in DA for a considerable while. It is only when there are no more than 10% of the neurones surviving any longer that the therapeutic efficacy of these drugs begins to decline, and one observes more side-effects and wider diurnal fluctuations. Although in theory the hypersensitivity of the post-synaptic receptors might be expected to compensate for the loss of neurons, in practice giving more drugs leads to a fall in sensitivity. In order to obtain the best effect with anti-Parkinsonian drugs, one should try to keep the dose as low as possible. As the disease progresses one often finds that one achieves greater success by lowering the dose than by raising it.

There is also a deficiency of NA in Parkinson's disease, and the major metabolite of NA, 3-methoxy-4-hydroxyphenethylene glycol (MHPG) is reduced in the basal ganglia and in the limbic system. Degeneration of the nerve-cells in the locus caeruleus leads to a loss of the supply of NA which is normally distributed to other brain regions. There is a dorsal noradrenergic tract which stimulates the nigro-striatal system. A ventral tract leads to the hypothalamus. DA-agonists are frequently adrenergic antagonists and therefore very often produce a marked diminution of the systolic blood pressure.

In many patients with Parkinson's disease there is a 40–50% reduction of 5-HT in many brain regions, but this reduction does not tend to get worse with the progression of the disease. Serotonergic neurones of the raphe nuclei innervate the striatum, among other regions, and there inhibit the dopaminergic tone. Ths is why L-tryptophan or 5-hydroxytryptophan very often accentuate akinesia, while adding tryptophan to an ongoing therapy with L-Dopa improves the patient's mood.

Neuropeptides

There seems to be no connection between metenkephalin and the destruction of dopaminergic neurones in the caudate nucleus, putamen and the nucleus accumbens. In the substantia nigra, on the other hand, the loss of DA seems to be correlated with the reduction in metenkephalin concentration.

GABA

There are frequently changes in GABA and in glutamate decarboxylase (GAD) in the substantia nigra and in the basal ganglia.

Acetylcholine

There is a deficiency of acetylcholine in the globus pallidus, in the putamen and in the frontal cortex (and this deficiency may possibly be contributed in part by anti-cholinergic treatment). While the cholinergic interneurons in the striatum are intact, there is a loss of cholinergic cell bodies in the nucleus basalis Meynert already apparent in Parkinson's disease without dementia. In patients with Parkinson's disease who are additionally suffering from dementia this loss amounts to more than 70%, and is accompanied by a loss of nerve terminals in the projection areas of the cerebral cortex.

Huntington's chorea

In this condition the small inter-neurones in the striatum become atrophied, but, in distinction to Parkinson's disease, the nigro-striatal DA-system is not affected.

The motor disturbances in Parkinson's disease

There are two phenomena to be kept in mind (see Table 1).

1. In Parkinson's disease we are not dealing exclusively with pathological changes in the synthesis and metabolism of a single neurotransmitter, most probably dopamine: many other neurotransmitters and neuropeptides are also affected. Thus we cannot account for most of the symptoms of the disease by attributing the symptoms to the specific change in concentration or activity of a single substance: rather it is the displacement of the normal correlations between the overall activities of individual neurotransmitters with one another. It was this recognition which led us to postulate the hypothesis of "the balance of neurotransmitters as a prerequisite for normal behaviour".

2. Although the idea that each neurotransmitter governs the same functions in the whole organism is beguiling, it is wide of the mark. For example while DA is unquestionably *the* neurotransmitter of extrapyramidal motor activity, however in the various centres of the hypothalamus DA participates in the regulation of sexual behaviour, in the expression of hunger-satiety feelings and in various other functions. 5-HT may be regarded as *the* sleep-promoting agent, but it also governs thermo-regulation. Substance P and enkephalin are spinal cord neurotransmitters for the ascending pain fibres, but in the substantia nigra substance P acts by modulating DA release. We can therefore see that the precise effect of a neurotransmitter varies according to the site of its action. This polyvalent effect of the individual neurotransmitters is the main reason

Table 1. Synopsis of the most important biochemical findings in Parkinson's disease

	Changes in Parkinson's disease (advanced stage) compared with normal ageing
Dopamine	↓↓↓
Homovanillic acid	↓↓
Tyrosine hydroxylase	↓↓↓
Biopterin	↓↓
Aromatic amino acid (dopa) decarboxylase	↓ (↓) =
Monoamine oxidase	=
Catechol-O-methyltransferase	=
cyclic-AMP-dependent protein kinase	=
D-1 Receptors	↑ ↓ =
D-2 Receptors	↑ ↓ =
Noradrenaline	↓ (↑)
Dopamine-β-hydroxylase	↓ (↓)
Phenylethanolamine-N-methyltransferase	↓
Serotonin	↓ (↓)
5-Hydroxytryptophan decarboxylase	↓ =
5-Hydroxyindole acetic acid	↓ (↓)
Serotonin receptors (5 HT-1 and 2)	↓ =
γ-Aminobutyric acid (GABA)	↓ (↓)
Glutamate dehydrogenase	↓ (↓)
GABA receptors	↓ (↓)
Substance P	↓ (↓)
Leu-Enkephalin binding	↓ = ↑

↓↓↓ greatly diminished; ↓↓ moderately; ↓ slightly; = unchanged; ↑ increased.
From: Birkmayer W, Riederer P (1985) Die Parkinson-Krankheit, 2nd edn. Springer, Wien New York.

for the multiplicity of symptoms in patients with Parkinson's disease.

The major distinction is between "positive" and "negative" symptoms". Tremor and rigidity are classical positive symptoms, whereas akinesia is classically a negative symptom, and it is the only symptom which is improved by the administration of L-Dopa (the DA-precursor), or which, if the drug is given in the early stages of the disease, may in certain cases completely disappear. Tremor and rigidity, on the other hand, are scarcely improved by L-Dopa. The way to treat tremor is with anti-cholinergic or anti-serotonergic drugs, which may offer some benefit: this supports the conclusion that tremor is caused by a relative excess of cholinergic or serotonergic activity, and that symptoms arise from a loss of the biochemical balance. The upswing of resting tremor that accompanies a change of mood in a situation where the patient is subjected to emotional stress is a sign of how closely the dopaminergic and noradrenergic activities are intertwined. Another example is that of the patient with Parkinson's disease who is severely restricted in his ability to move, but who wants to cross a very busy road. The strong apprehension he feels overcomes the restriction and so improves his motor capacity – but only until the danger of getting run over is past. This elevation of motor performance by psychological stimulation is a little like the way a culture of protozoa erupts into activity after chemical stimulation. The exact opposite of this potentiation of the ability to move by the threat of danger is the "dead-still" reflex (or "playing possum", because the opposum has the reputation of having perfected the feigning of death in response to the threat of danger). When we consider the symptoms in patients with Parkinson's disease, the dead-still reflex can be observed in the so-called "freezing" phenomenon. As a result of apprehension or anxiety the patient is completely incapable of moving his legs and of initiating walking motions. Freezing can just as easily occur in the confines of the patient's home because he is afraid of bumping into things or of falling over, as in the street, where he is afraid of crossing the road. Thus we see that anxiety, as an emotional NA-activation, is just as likely to stimulate a dopaminergic surge of movement as a dopaminergic blockade of mobility (dead-still reflex). Again, while 5-HT arrests tremor during deep sleep, during REM-sleep (the NA-phase) tremor returns.

Rigidity presents a situation where there is a simultaneous tonic innervation of agonists and antagonists. The gamma loop represents the peripheral system for regulation of tone. Tracts of nerve fibres go out from the anterior horn of the spinal cord: the large alpha-cells are responsible for the dynamic innervation of the extremities, the small alpha-cells for the tonic innervation of the muscles of the trunk, and besides these there are also the gamma-cells which send nerve impulses to the sensitive muscle spindles. Gamma-1 fibres stimulate nuclear bundle fibres, whose stimulation leads, via 1-A-fibres of the spinal cord, to the small alpha-cells of the anterior horn and maintain the dynamic postural tone. Gamma-2 fibres, which stimulate the nuclear chain fibres, stimulate the static component of muscle tone via afferent II-tracts. This regulates the innervation responsible for keeping posture upright against the forces of gravity, while the dynamic tone is responsible for maintaining the tonic innervation needed for adaptation to various changes in posture. In Parkinson's disease both these information systems are inadequate. Static tone is reduced, and patients therefore tend to stoop and are incapable of standing upright for any length of time; the dynamic tone likewise shows inadequacies: if one pushes a Parkinsonian patient in the chest, he usually falls over backwards, because he is unable to adjust his dynamic tone to compensate with the required change in posture. Since rigidity can be reasonably well compensated with anticholinergic drugs, and is made worse with physostygmine, one must assume that rigidity is caused by a loss in the balance between DA and ACh in favour of the latter.

Akinesia is the consequence of the patient's inability to convert a potential energy of movement into actual kinetic motion. Whereas paralysis is due to a structural, and in the CNS mostly incurable lesion, akinesia is an inability to move which, at least in the early stages of the disease, can be completely lifted by administration of L-Dopa. We can compare the striatum with a flat battery which only needs re-charging to deliver

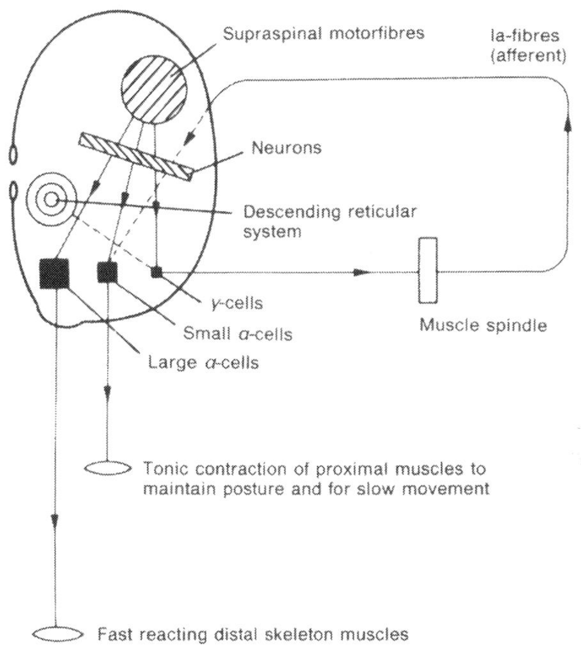

Fig. 12. Diagram of the gamma-loop. Supraspinal motor pathways have an inhibitory effect on the gamma cells, while the reticular system has a stimulatory effect. Lesions of the pyramidal tract increase the activity of the gamma loop ("gamma spasticity"). Loss of the reticular stimulation leads to an alpha-activation

current again. A characteristic feature of the disturbed motor abilities of patients with Parkinson's disease is that they can neither intend nor excute sudden movements of maximal intensity, and neither can they bring any motion to a smooth stop. They also have difficulty in initiating any movements that act against the forces of gravity. The salient feature of the akinetic movement disorder lies in the inability of the body to carry out the first instruction of the mind: turning over in bed, getting out of a chair, taking the first step, in all of these the interplay between intention and performance is particularly disturbed in Parkinsonian patients. In healthy people the mental and emotional elements can increase or decrease the quality and quantity of movements at will: in states of fear (surge of NA) one can run faster, but in sudden terror one can also be rooted to the spot. The Parkinsonian patient exhibits both these possibilities of action, but in extreme, exaggerated form. Fear may block his movement, but there is also the phenomenon of paradoxical kinesia where the patient is galvanised into action by fear, and can run over long stretches although normally he cannot even get up out of his chair. This can be explained by the fact that a surge of NA can also elicit a release of DA, but of course for only a brief period, for only as long as DA stays available.

The surge of NA reaches the substantia nigra via two pathways: a dorsal bundle of nerves promotes and a ventral bundle inhibits the DA-system, and so a surge of NA released from the locus caeruleus may either produce a dynamic response (fight or flight) or an inhibition (playing possum).

The loss of the normal rhythmic swinging gait forces the patient with Parkinson's disease to walk with short steps that minimize any risk – risk of stumbling, of tripping, of falling. And difficult as starting may be, stopping is just as hard. The propulsion, i.e. the acceleration of the succession of steps, eventually causes the patient to pitch forward, unless he can reach the safety of some nearby piece of furniture to clutch.

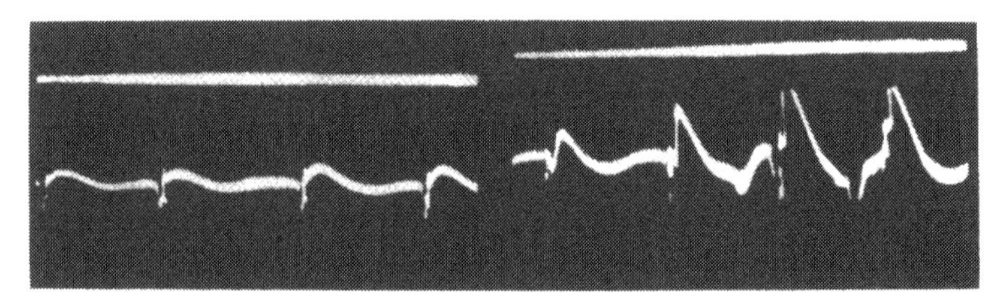

Fig. 13. T-Reflex in a patient with Parkinson's disease. Left: untreated; right: 30 minutes after intravenous administration of 50 mg L-Dopa. [From: Danielczyk W, Department of Neurology, Lainz Hospital, Vienna]

Walking with one or two sticks is more like the motion of a quadruped, and makes the patient feel more secure, but patients accept walking with sticks only very reluctantly, because they dislike making their crippled state so obvious. Speech presents much the same difficulties. Starting, getting the first word out, is inhibited, and then, just as with walking, stopping is just as hard. Speech goes gallopping ahead, faster and faster, and harder and harder to understand. A smooth start and a smooth ending are alike inaccessible for all movements, speech included, and the same applies just as much to writing.

The mask-like face of the patient is an akinesia of the facial muscles preventing him from expressing his feelings and emotions. As a rule the patients stare straight ahead and appear to be completely uninvolved, but if they are spoken to, then some expression flits across their features, i.e. an emotional arousal (NA) leads to a reaction producing a DA response, perhaps a smile or a turn of the head. The patient usually has a rather stooped posture, and those around him will tend to keep reminding him "do stand up straight . . ." The patient does briefly intend to straighten his back, but he cannot keep it straight for long, and this can be blamed on the diminished activity of the gamma-loop. This also makes walking upright and the associated motions such as walking and running difficult. Sitting down, on the other hand, often leaves the activities of hands and legs unaffected, so that patients can continue driving longer than they can walk.

A particularly characteristic feature of patients with Parkinson's disease is their sensitivity to changes in the weather. The adaptation of the organism to climatic changes is mediated via neurotransmitters. Warm weather and high atmospheric pressure lead to a release of NA, while cold rainy weather releases 5-HT. In older people there is an inherent weaking of adaptability, which can be related to a decreased availability of neurotransmitters. Patients with Parkinson's disease find hot weather particularly trying, and feel much more at ease in a cool environment. They also reflect the diurnal rhythm: most patients are more lively in the morning, and after lunch they feel more lethargic and sluggish. However, within the framework of depressive phases there is a relative inactivity in the mornings, which only improves in the afternoon. Although it might have been tempting to explain fluctuations, which in extreme cases went as far as a total "off"-blockade, by changes in the plasma dopa level, this would have been erroneous. The motor behaviour of the patient does not depend on the plasma dopa level, but on the availability of DA in the neurone. An off-phase is the expression of a disturbed DA-utilisation, and the most satisfactory way for the patient and his doctor to deal with it is to wait until such a dopaminergic blockade switches over to an active phase.

Autonomic disturbances in Parkinson's disease

In addition to the motor disturbances, and running alongside them, there are also positive and negative symptoms in autonomic functions. The most frequent are excessive salivation, seborrhoea, sudden sweats, flushing, fever, and oedema of the legs. Whereas the salivation problem can be improved with anticholinergic drugs, flushing, oedema and fevers are best treated with L-tryptophan.

Another system that originates autonomically is compulsive slimming, which is related to teenage anorexia nervosa. Hyperactivity of the dopaminergic system in the centres of the hypothalamus that are responsible for eating behaviour are responsible for this problem. This DA-hyperactivity depresses the serotonergic systems which are associated with the dopaminergic ones in the regulation of hunger and satiety. The Parkinsonian anorexia may involve a similar biogenic amine mechanism. A relative rise in catecholaminergic activity and a corresponding decrease in serotonergic activity occurring to different extents in different brain regions and in different cybernetic relationships may provoke insomnia, constipation and also the anorexia. However, Parkinsonian patients generally have nothing wrong with their appetite, on the contrary, they are always eating and at times massively. These clinical observations point to fluctuations in eating behaviour which may be related to the progressive degeneration and the resulting increasingly unbalanced relationships between the neurotransmitters.

Besides the symptoms already described, the patients frequently complain of burning feet, itching, tingling, restless legs, hot sensations and tightness of the chest. These symptoms of autonomic disturbance, accompanied with ill temper, are often found in masked depressions, but they are also often apparent in Parkinson's disease, and point to a regional loss of biochemical equilibrium in the central nervous system. This central biochemical disturbance increases progressively, even though the peripheral disturbances are only minor, and this is further clinical proof of the influence of the central control mechanisms on the function of the peripheral organs.

These autonomic and subjective disturbances shade over into depressive phases, and these are by no means rare in Parkinson's disease. In 11% of patients depressive phases even appear some years before the appearance of Parkinsonian symptoms. Post mortem biochemical analyses on patients with depression have shown biogenic amine depletions similar to those found in Parkinson's disease, admittedly mainly in the limbic regions, but also in the red nucleus, which may have contributed to the poor posture of the depressive patients as in Parkinsonian patients. The distinction between the depressive phase of Parkinson's disease and a true endogenous depression lies in the fact that the latter show much more severe mood disturbances and last much longer, while those of Parkinson's disease are less severe and of shorter duration. Beside the autonomic and affective problems of patients with Parkinson's disease there are also intellectual deficiencies, and these take two main forms: bradyphrenia, a slowing of thought processes accompanied by a loss in concentration and a shortened attention-span; and true dementia.

Bradyphrenia is not a genuine loss of intellectual ability, but rather a kind of mental akinesia. The speed of mental activity, as well as the ability to make rapid decisions and judgements is slowed down, although there is no reduction in intellectual capacity. This specific impairment of intellectual function is probably again the result of a deficiency of DA which may be cybernetically referred to a different system from that responsible for motor akinesia. Reduced dopaminergic activity in the mesolimbic/mesocortical DA-system arising from the ventral tegmental area seems to be involved in the pathophysiology of bradyphrenia. Since there is additionally a deficiency of NA-activity in the reticular formation and in cortical areas, the diminished arousal activity impairs the tempo and capacity of thought.

In our experience the prevalence of dementia is on the increase. The exact cause of this dementia has not been completely established. One possible explanation is that thanks to modern methods of treatment patients with Parkinson's disease survive longer, and so more of them are likely to contract dementia. There is a good deal of argument about the assumption that treatment with anticholinergic drugs over a period of years may be a causative factor in this dementia. Presenile Alzheimer's disease is particularly characterised by a massive loss of memory, but this very impressive clinical sign is not a feature of senile failing of intellectual function generally. It may therefore be more to the point to take the view that the Alzheimer-type of symptoms so frequently seen within the general framework of Parkinson's disease is more likely to be one caused by a) a loss of over 70% of cholinergic neurons of the nucleus basalis Meynert and probably also of cholinergic projections arising from the medial septum and the area of Broca, and b) by features of the so-called "Dopa psychosis". In this condition there is over-stimulation of the mesolimbic DA-system, or a major depletion of e.g. NA and 5-HT from their specific neurones, and these effects produce a confused hallucinatory syndrome rather like a kind of exogenous drug reaction. Anticholinergic drugs, used as anti-Parkinsonian medication, inhibit cholinergic activity not only in the striatum but in all brain regions, including the cortex. Whether there is a depletion of acetylcholine caused by excessive concentrations of DA has not, up till now, been evaluated. It cannot, however, be ruled out that other transmitters beside ACh may play an additional role in the pathogenesis of dementia in Parkinson's disease. The biochemical deficiencies in Parkinson's disease are also not restricted to the striatum, but affect the whole midbrain and limbic system. The substantia nigra represents a sort of biochemical "distributor head" between motor, affective and intellectual functions, and consequently we repeatedly observe these imbalances during the course of Parkinson's disease, expressing themselves in the form of depressive, dementive, pseudo-neurotic and psychopathic disturbances.

We have never observed any addiction to alcohol, abuse of medicines or of hard drugs in Parkinsonian patients. The multi-factorial manifestations of the disorder in all biological systems result in the patient's inability to adapt not only to stresses in the overall environment (e.g. hypersensitivity to changes in the weather), but also to stresses in the social milieu. The Parkinsonian patient is just not capable of calmly absorbing and coping with any stress by feedback mechanisms, and this stamps his whole behaviour. His inadequate adaptability vis-a-vis other patients or members of this family is such that no other neurological illness presents greater problems in nursing and management than Parkinson's disease. There are no biochemically directed forms of treatment that can deal with this maladaption, because specifically directed drugs are likely to lead to disturbances in other biological systems. Antidepressive drugs can damp down the instability of biochemical regulation. Mastery over these modes of adaptation does enlarge the patient's "personal space" or biological territory, and so extends the psychological distance between patients in an institution or between a patient and members of his family at home, reduces areas of interpersonal friction and so diminishes social stresses. These universal principles for diminishing interpersonal stress, for extending people's "personal territories", do not hold only for Parkinsonian patients, but are generally applicable to mankind as a whole.

The treatment of Parkinson's disease

Since we are putting forward the principle that the way to treat all neurological disorders is to try to neutralise the biochemical imbalance by specific replacement therapy, it follows that for Parkinson's disease the rational objective and the most important aspect of treatment is the replacement therapy using L-dopa. However, at the time when symptoms first appear, only 30% of functional DA-neurones are left, and so the dosage of L-dopa must be carefully adjusted to just the amount that the still remaining neurones are able to accept: too little and there is insufficient effect, too much and side-effects will appear.

In the striatum there is an equilibrium between cholinergic and dopaminergic activities, which is normally somewhat in favour of DA. Where the dopaminergic activity is deficient, ideally, administration of L-dopa restores the biochemical balance. If we give too much, then hyperkinesias appear, in the form of overshooting movements. Our normal emotional behaviour is likewise dependent on the biochemical balance between, among others, DA, NA, ACh, GABA and 5-HT. But an excessive dosage of L-dopa distorts the normal cybernetically determined mutual correlation of these neurotransmitters, and this can result in the appearance of side-effects. The sleep disturbances that may appear during treatment with L-dopa, and which are caused by catecholaminergic hyperactivity or by the depletion of serotonergic neurones by the L-dopa that was administered, are thus elicited by hypofunction of the sleep-transmitter 5-HT. In other words, L-dopa medication is only a therapeutic measure when the dosage is optimal. Keeping the dosage down therefore on the one hand avoids unpleasant side-effects, and on the other hand it prevents too rapid a progression of the degenerative process.

Thus the objective of modern treatment of Parkinson's disease is supplementation with various additional drugs whose function is to improve the effectiveness of L-dopa without any increase in its dosage. The first of these additional drugs was benserazide, which is an inhibitor of the peripheral aromatic amino acid decarboxylase. This decarboxylase synthesises the neurotransmitter DA from L-dopa, both in the CNS and in the periphery. In the CNS this synthesis is desirable, but in the periphery the excess DA produced causes undesirable side-effects (nausea, vomiting, disturbances of the circulation). Benserazide and carbidopa block the peripheral decarboxylase, and via improved brain dopa-uptake they permit the entry of about six times as much DA into the CNS as without them, and this allows a substantial reduction in the dose of L-dopa. While 100 mg of pure L-dopa, given orally, scarcely produces any motor effects, madopar (100 mg L-dopa plus 25 mg benserazide) is a well-tolerated and effective combination.

A further essential additive is selegiline. In earlier chapters we saw that the enzyme MAO breaks down certain biogenic amines, and that it exists in the A- and the B-forms. Type A breaks down NA (and 5-HT), while type B breaks down DA, (5-HT) and phenylethylamine. Selegiline blocks the activity of type B, and is an effective supplementary drug. By permitting the stockpiling of DA in its storage sites it substantially improves akinesia, it contributes to a significant extension of life-expectancy and it reduces the incidence of side-effects. It also diminishes fluctuations in the realm of motor activity: the characteristic variations that occur during the course of the disease, and which are caused by excessive L-dopa dosage, are substantially reduced by selegiline.

Further additional drugs – direct stimulators of the post-synaptic receptors – are bromocriptine and lisuride. Direct stimulation of these receptors improves akinesia in spite of a reduced dosage of L-dopa. These additional drugs give positive results even in the later stages of the disease, when the amount of DA which can be synthesised and utilised is no longer sufficient. A low dosage is particularly important for therapeutic success in the later phases of the disease, because the fewer DA-neurones are left, the more the incidence of side-effects goes up. The most significant side-effect of dopaminergic agonists, in our experience, is orthostatic hypotension: when the patient gets up his systolic blood-pressure may drop to as low as 70 mm Hg, and this usually causes dizziness, and more rarely, collapse. Dopaminergic agonists exert their optimal effect only when L-dopa is administered as well. They seem to be able significantly to reduce dyskinesias when given before the start of combined L-dopa treatment (L-dopa together with peripheral decarboxylase inhibitors).

Another additional drug is amantadine, introduced by R. Schwab in the U.S.A. in 1969. Most doctors use it in the early stages of the disease. It gives clinical improvement in less severe cases, but the improvement is much less pronounced in the more severe cases. Although this drug has very few side-effects, its effectiveness is only modest. Wesemann was able to show that its mode of action is to change membrane fluidity. On the one hand this facilitates the presynaptic release of DA, and on the other it improves the post-synaptic response, which is particularly significant in akinetic crises. The modest effect of this drug in comparison with L-dopa is understandable because it scarcely interacts with neurotransmitter metabolism.

The oldest drugs used to treat Parkinson's disease are the anticholinergic drugs. They have only a slight kinetic effect, but they improve the balance between striatal cholinergic and dopaminergic activities. They are effective against excessive salivation, seborrhoea, but less effective against sudden sweats and only occasionally also against tremor. Recently there has been some argument about whether long-term use of anticholinergic drugs may not tend to induce dementia.

A very unpleasant symptom in the realm of autonomic functions is constipation. It appears above all when treatment is solely with L-dopa, and is produced via catecholaminergic and serotonergic mechanisms of the intestines, so one might assume that supplementation with tryptophan (as 5-HT precursor) would stimulate peristalsis and improve the constipation. However, since benserazide, which is combined with L-dopa in modern treatment, also blocks the decarboxylation of tryptophan, its administration has little success.

Since Parkinson's disease is characterised not only by a deficiency of DA but also by a considerable lack of NA, it is hardly surprising that in the later stages of the disease orthostatic hypotonia should appear. Patients complain of dizziness and uncertain gait: they sway about as if walking on clouds. This hypotonia is generally made worse by changing from sitting down to standing up, and systolic pressure values of 50 mm Hg are not unusual. Not infrequently such a low value results in collapse with loss of consciousness, usually lasting only a few minutes and passing off without any psychological effects. These orthostatic effects are particularly noticeable after bromocriptine or lisuride, and are scarcely improved with peripheral NA-stimulation (e.g. etilefrine). We have however been able to see a noticeable correction of the orthostatic fall in blood pressure with the administration of dihydroxyphenylserine (DOPS, the immediate amino acid precursor of NA) either orally or intravenously.

Antidepressive drugs with a dynamic effect (dibenzipine, three times daily; melitracen 25 mg twice daily) occasionally also show an NA-potentiating effect, and these drugs are of course also successful against the occasional periods of intermittent depression. In cases of dispiriting depression, I have recently been giving 10 mg of selegiline plus 250 mg of phenylalanine in the form of an infusion, and in the evenings a sedative (amitriptyline or doxepin in doses of 10–25 mg). Oral administration of selegiline, 5 mg twice daily plus twice 250 mg phenylalanine is a useful combination in cases with less severe symptoms. In patients who are anxious and agitated one can give infusions of 250 mg tryptophan with 20–30 mg selegiline. The beneficial effects start to show themselves after only a few days. Incidentally, since we include anorexia nervosa with the group of masked depressions, we also administer these particular drug regimes to those patients.

The intellectual slowing in Parkinsonian patients is best handled by an initial increase in L-dopa, and then adding in 3000 mg of piracetam in the form of a drinking ampoule or of an infusion. A tablet of pyritinol twice a day, or perhaps cerebrolysin, have a beneficial effect, both subjectively and in fact, on disturbances of recall. Unfortunately, however, there are as yet no rational modes of treatment of authentic Alzheimer's disease.

A synopsis of treatment for Parkinson's disease

There are two basic principles:

1. The precursors have to be present in the plasma in sufficient concentration, because without the availability of an adequate supply of precursor in the bloodstream there will not be a satisfactory rate of entry into the brain or synthesis of the neurotransmitter in the neurone.

2. The supply of precursor made available to the neuron must be utilised, i.e. adequate capability for converting it to the appropriate neurotransmitter (DA, NA or 5-HT) must exist within the neuron.

Availability and conversion-capability can be regarded as the two essential processes of neurotransmitter synthesis. Availability is relatively easy to attain through a balance in dosage leading to more constant plasma levels, but at the moment we have no means of affecting the ability of the neuron to convert precursor to neurotransmitter. The application of the amino acid L-dopa was one of the golden moments of modern neurology. However, an essential consequence of this application is that as progressive degeneration diminishes the number of remaining functional neurons, so we are forced to reduce the dose of L-dopa. If we try to ignore this necessity then side-effects will appear. The classical situation of such side effects is the "dopa-psychosis". Because of the progressive loss of striatal neurons, the dopa that is administered can no longer be completely converted to DA and put into storage, and so there is DA overstimulation in extrastriatal regions, or in certain circumstances a displacement of 5-HT and NA from their stores within the neurones. The clinical effects of this are anxiety, insomnia, confusion, hallucinations and delusions. If now one adds L-tryptophan or 5-hydroxytryptophan together with 3,4-dihydroxyphenylserine (DOPS), then the growing availability of these precursors results in a restoration of the balance of respective neurotransmitter function, and the clinical symptoms disappear.

This mechanism was the basis for the formulation in 1972 of our hypothesis concerning the "balance of neurotransmitters as the essential requirement for our normal functioning and behaviour". The occurrence of a transient psychosis can be observed not only with L-dopa, but also with bromocriptine, with amantadine, with selegiline and also with anticholinergic drugs. Because such occurrences depend on dosage levels, the principle of an optimal therapy must be to use different drugs in different specific biochemical areas, in order to get as close as possible to the physiologically balanced effect while at the same time maintaining an effective dynamic equilibrium.

The second most important side-effects are the hyperkinesias. There is normally an equilibrium between DA and ACh in the striatum. In Parkinson's disease the DA-level declines, leading to akinesia. If addition of L-dopa pushes the DA-level up beyond its normal point, then hyperkinesia occurs, and administering anticholinergic drugs is no help either. These hyperkinesias can only be damped down by a reduction of the L-dopa dosage. In extreme cases of hyperkinesia one can give a neuroleptic drug, and the one of choice is tiaprid. Painful cramps, mostly in the legs, may occur, particularly at night, and are most easily dealt with by simple excercises which the patient can do either in bed or during walks. Reducing the evening dose of L-dopa also gets rid of the cramps. Recalcitrant cases usually respond to 3–6 mg of bromazepan in the evening.

Sleep disorders involving extra-vivid dreams occur in 8% of our patients; the dopa treatment tends to lead to a damping down of the serotonergic neurons of the mid-brain, which may cause a loss of sleep. The relative preponderance of NA compared to 5-HT promotes REM-sleep with its dream states. This is adequately treated with L-tryptophan, between 500 and 1000 mg at night, or with a sedative antidepressant drug such as amitriptyline (10–25 mg at night).

In the advanced stages of the disease patients complain of dizziness. There are three main forms of dizziness. The first of these is the genuine Ménière's attack, often set off by turning over in bed. The cause may be a lesion in the vertebrae of the neck region of the spine, and on turning one of the vertebral arteries may get occluded. Second, and most common, is the dizziness caused by orthostatic hypotension. The inherently low Parkinsonian blood pressure (in general 110 over 80) drops to 50 over 40 on standing up, and this leads to a collapse, with loss of consciousness, particularly on prolonged standing. If one looks at a patient one can see right away from the upward-directed gaze and the rolling of the eyes, "... he's going to fall unless we sit him down or put him to bed ..." This orthostatic dizziness can often also be triggered by too much bromocriptine, and dropping the L-dopa dose does not help, neither do peripheral NA-stimulator drugs like etilefrin etc. In the less severe cases one can achieve some success with NA-stimulating antidepressive drugs (e.g. dibenzipine, 80 mg three times a day). The orthostatic hypotension is probably caused by destruction of the locus caeruleus, so in some cases DOPS (200–500 mg intravenously, or 200 mg three times a day orally) can raise the blood pressure, and can prevent the precipitous drop in blood pressure on standing up. The third form of dizziness is associated with nutritional disturbances.

An analysis of the time of death during the course of the daily rhythm has shown that patients with Parkinson's disease mostly die at night or in the early hours of the morning, and this is in contrast to a control group of patients with other neurological diseases, of whom a high percentage died during the daytime hours. This too is evidence that the biological homeostasis of neurotransmitters persists to the very end, and that life in Parkinsonian patients is most likely to flicker out in a parasympathetic phase.

Depression

Our involvement with depression spans a period of over thirty years of observations, investigations and discoveries which I (WB) have made in patients with Parkinson's disease.

About a quarter of all patients with Parkinson's disease suffer depressive periods during the course of the disease, or even in many cases (11%) before its onset. These depressive phases are connected with the disturbances in the biochemical metabolism of affected patients. In patients with endogenous depression one keeps coming across members of their families who also suffer or used to suffer from depressions, but with Parkinsonian patients there are many families without any incidence of depressions: in other words the depression of a patient with Parkinson's disease is not hereditary or familial, but is a manifestation of the disease, and is caused by a loss of the neurotransmitter balance. Parkinsonian depressions are usually of lesser severity and shorter duration.

In the foreground is the affective disturbance, with the patient feeling out-of-sorts and suffering from a lack of emotional drive, loss of appetite, of sleep, of mood, and of the daily rhythm of life. Depressed patients with Parkinson's disease tend to be seized-up in the morning, as far as movement is concerned, and it is only in the afternoon and evening that their mobility returns to normal. As a consequence they naturally tend to attribute their improved mood to their improved mobility, but the two are not necessarily cause and effect. There are countless Parkinsonian patients whose therapeutic state is well-adjusted and who show good mobility in the mornings, but who, in a depressive phase, are highly dissatisfied in the mornings, querulous, negative in their attitude to their medicines and towards all the other efforts on the part of their clinicians. Once one has experienced the course of the disease in a number of patients, one soon recognises these scratchy moods, and can immediately diagnose "depression" and introduce antidepressive treatment, and usually it is then only a matter of a few weeks before those caring for the patients report them to have become more tractable and contented.

Biochemical analyses of the brains of our depressive patients who have died, essentially show much the same sort of picture as in Parkinson's disease, only the abnormalities are less marked. Because the manifestation of symptoms is less pronounced, less powerful antidepressant drugs are found to be quite effective. The usual depressive changes only require dibenzipine, 40–80 mg in the morning, and then 10 mg of amitriptyline at night, but these drugs have to be maintained for several months. Side-effects are very rare, with one exception: patients do complain about putting on weight. This effect, which signals that the recovery phase, which is stimulated by 5-HT, has arrived, is also seen in patients with Parkinson's disease. In true depressions we were able to see post-mortem alterations of the neurotransmitter concentrations compared with controls. The characteristic feature of these was not only the diminished concentrations of DA, NA and 5-HT, but also the distorted correlations between them. These were quite variable, and these variations may well be responsible for the appearance of individual symptoms. Tranylcypromine and selegiline at dosages of 10–15 mg block the intraneuronal degradation of the neurotransmitters (NA and 5-HT, and DA and phenylethylamine respectively). These two compounds are, in our experience, rapid-acting and effective antidepressants. We do not as yet have any drugs which effectively stimulate neurotransmitter-synthesizing enzymes, and so blocking the activity of the degradative enzymes is the more effective approach.

Since it is only rarely that we can carry out post mortem analyses on the brains of deceased depressive patients, we have tried over the last ten years to analyse the neurotransmitter precursors tyrosine and tryptophan in the plasma, and their metabolic end-products (HVA, VMA and 5-HIAA) in the urines, of about 1000 depressed patients. We found deviations from normal control values for plasma tyrosine and tryptophan concentrations. At the same time we analysed early morning urine for HVA, VMA and 5-HIAA. We used early morning urine because preliminary investigations had shown that in depression it was the early morning specimens that most clearly demonstrated the reduced HVA and VMA excretion, i.e. that the metabolism of DA to HVA and of NA to VMA was slowed down.

The way that these reduced urinary concentrations parallel the inactivity in the morning is very striking, but it is not possible to prove a direct connection between insomnia and reduced 5-HIAA excretion, or between an increased tyrosine turnover to VMA and periodic anxiety. Biochemical analyses of peripheral metabolism therefore only give us evidence of depressive occurrences averaged over a large number of patients, but they are more likely in the somatic type of depression (masked depression).

We have already mentioned that a specific neurotransmitter may not exert identical functions in all regions where it may be active. Thus an increased NA-release in the amygdala may cause aggressive behaviour, while in the reticular formation it may lead to greater alertness. However, one thing we can definitely state, and that is that in no depressive patient, and above all in no patient with masked depression, where we analysed all these five parameters, did we find them all to lie within the normal ranges of statistical deviation. And so, whenever we came across so-called borderline cases that fell between depression and schizophrenia, if the analytical values fell more into the categories appropriate to the depressive syndrome, then the patient was given antidepressive treatment, regardless of his clinical picture. If, on the other hand, the clinical picture of anxiety, of disturbances of arousal, of insomnia, and of remission of symptoms in the evenings, pointed in the direction of depression but the biochemical analyses gave normal values, then treatment with neuroleptic drugs was instituted, as more likely to be successful than antidepressive therapy.

When one has experience of a large number of patients, then one is in any case left with the impression that these borderline cases, which exhibit a clinical picture of both depressive and schizophrenic symptoms, are appearing more frequently nowadays. I (WB) have been observing psychotic patients for fifty years, but the phenomenon of these more frequent compounded cases is new.

We believe that the cause may lie in long-term – and possible not rationally designed – drug therapy. If one administers antidepressant drugs to patients with latent Bleuler's syndrome, one nudges them towards schizophrenia; if one administers neuroleptic drugs to patients with depression, one frequently obtains accentuated depressive symptoms: this clinical experience already points to the conclusion that we have to regard depression as a disturbance of the balance between the neurotransmitters. A neuroleptic blocks certain receptors, and thus diminishes the effect on specific neurotransmitters, and aggravates the depression through reinforcing the "negative" symptoms. On the other hand, when one gives an antidepressive drug during a schizophrenic episode, the episode becomes dramatically more acute, because amongst other effects the re-uptake blockade potentiates the action of the neurotransmitter by reinforcement of the "positive" symptoms.

This brings us to the question of what causes depression. The disturbances of the biochemical equilibrium which we have presented are responsible for eliciting the depressive symptoms; but we do not know what induces the disturbance of the equilibrium between the various neurotransmitters.

Normally distortions of biochemical functions are evened out by negative feedback mechanisms. Feedback regulation of an anxiety symptom (including agitation) can eventually lead to exhaustion of the catecholamine stores and so to a relative preponderance of serotonergic activity, resulting in sleep as a compensatory process. This kind of mechanism could often be observed during the war, when soldiers anxiously awaiting an attack would suddenly fall asleep.

Such feedback processes are disturbed in Parkinson's disease, where an excessive dosage of L-dopa may produce hyperkinesia. In a healthy person there is an efficient negative feedback control which rapidly deals with the excess of DA. But if the functional pathway has a missing step, then the correction cannot take place. The possible sites of such gaps are for example at the nerve-endings (intraneuronal coupling) or at the receptor (via interneuronal coupling). If the specific receptor is not coming into play, then feedback cannot be applied. Another possible explanation for inadequate compensation might be that there is a functional deficit in the compensating neurotransmitter itself. Without NA there can be no arousal reaction. In Parkinson's disease the progressive neuronal degeneration can eventually shut off every possibility of such compensations.

In depressions there is no underlying degenerative process at work: rather there are neuronal disturbances. These functional fluctuations disturb the biochemical balance and thus lead to the diverse symptoms of disturbed feelings and behaviour.

This section is not intended to be a systematic report of the various clinical forms of depression: we wish rather to reveal the way the eliciting transmitters participate. The fact that there is a world-wide increase in depressive illness is evident: it indicates that the increased stresses of modern life have placed heavier burdens on the neurotransmitter systems. We can nowadays already see depressions at puberty, during and after pregnancy, at the menopause, during the post-menopausal period and finally also in old age. This all points to the fact that at critical times in life, times that put increased demands on our biochemical systems, a syndrome of defective adaptability may appear, to which we just give the label "depression". In some classifications there is a category described as "neurotic depression", but in our view there is no such clinical entity. It represents an attempt on the part of a frustrated patient to overcome negative aspects of his or her life via neurotic mechanisms, and such neurotic modes of behaviour are so pronounced that the attending physician may completely overlook the basic underlying depression. While the depressions at these various stages of life, are usually capable of remission, there are some depressive disturbances, particularly in old age, that offer no such possibility. They include depression following severe influenza, after surgery involving prolonged anaesthesia, and also where arteriosclerosis causes a reduced blood flow. They may be described as "somatic" depressions, because they are caused by recognizable somatic defects.

A large number of rating scales have been devised to provide an exact, objective method of characterising the clinical syndrome (e.g. Hamilton, Zung, von Zerssen and many others). We use one which we have devised ourselves (see Table 2), which can be readily understood by every doctor and patient, which does not take long to score, and which provides adequate indications for a rationally designed treatment. The first group indicates six categories of symptoms particularly indicating unsatisfactory activity of catecholaminergic neurotransmitters. The next group, numbered from 7 to 11, points up symptoms indicating serotonergic insufficiency. No. 11, loss of libido, however, probably depends rather on a DA-deficiency in the corresponding sex-related centres. We know, for instance, that the septum pellucidum in the limbic system is responsible in part for the release of desire and for its satisfaction. The not infrequent symptom of amenorrhoea in depressed women is probably due to a biochemical defect which may well turn out to be connected with the hypothalamus. From there dopaminergic nerve tracts run to the pituitary, and thus influence its hormones. The effect on prolactin secretion is inhibitory, and the inhibiting factor may well be identical to DA. Prolactin acts mainly on the breasts and sex organs, and so here we have a classical example of neurotransmitter-

Table 2. Rating scale for depression (Professor W. Birkmayer)

Parameter to be rated	Score				
	0	1	2	3	4
1. Lack of enterprise	–	–	–	–	–
2. Lack of enjoyment	–	–	–	–	–
3. Lack of interest	–	–	–	–	–
4. Lack of initiative	–	–	–	–	–
5. Lack of concentration	–	–	–	–	–
6. Reduced work capacity	–	–	–	–	–
7. Loss of sleep	–	–	–	–	–
8. Loss of appetite	–	–	–	–	–
9. Loss of weight	–	–	–	–	–
10. Constipation	–	–	–	–	–
11. Loss of libido	–	–	–	–	–
12. Remission in the evening	–	–	–	–	–
13. Compulsive brooding	–	–	–	–	–
14. General pessimism	–	–	–	–	–
15. Self-reproach	–	–	–	–	–
16. Feelings of guilt	–	–	–	–	–
17. Anxiety	–	–	–	–	–
18. Suicidal tendencies	–	–	–	–	–
19. Hypochondriac ailments	–	–	–	–	–
20. Feelings of the futility of life	–	–	–	–	–

hormone coupling. The third section, numbers 12 to 20, includes depressive symptoms which we have been unable, in our present state of knowledge, to relate to any specific biochemical disturbance. The deciding factor for the remissions that occur in the evening is a more vigorous outpouring of neurotransmitters, one more in tune with requirements, and its greater vigour probably represents the crest of a wave within a diurnal fluctuation.

The "negative" symptoms in depression are a compulsive introspection, pessimistic notions, needless worries about the future, blaming oneself and feeling guilty, symptoms which are caused by a crucial inability to cope psychologically. Guilt feelings, which incidentally were part of the classical clinical picture of melancholia at the turn of the century, are scarcely mentioned much in these emotionally deprived times. They represent a consciously experienced reflection of biological inadequacy. Suicide, or suicidal thoughts or impulses represent, in the widest sense, a flight from mental turmoil into the tranquillity of death. Anxiety, the alarm signal which is biochemically sounded through prolonged NA-activation, is at the same time also the emergency reaction, and this reaction, which is supposed to lead to a biochemical adaptation, is missing in depression accompanied by suicide.

Heading 19, hypochondriac complaints, are mostly seen in the special case of the masked depressions. Here symptoms in the periphery of the organism become prominent, and this indicates a disturbance of peripheral as well as central neurotransmitter balance. These patients complain, for example, of stomach pains and keep trying to convince their doctor that their problem is gastritis, and not depression. They may well actually have an upset stomach, because in depression the previously described central or peripheral disturbances of the equilibrium between the neurotransmitter systems may well cause a lack of gastric acid secretion, a dryness of the mouth an insufficient secretion of digestive enzymes and also a sluggish peristalsis. Finally, in depression there is also a generally reduced muscular tone, which leads to a bent posture, and the affected patient lets his head droop. At the same time his voice is flat and unmodulated and he has an expressionless stare.

This compilation clearly demonstrates how disturbances in the co-ordination of these chemically active substances in the various regions of the brain-stem and limbic system are capable of provoking the most varied symptoms. The company of symptoms which muster on parade in depression, taken as a whole, is so characteristic, that depression is easily diagnosed, and parading alongside this company of symptoms is a corresponding company of neuronal defects. It is this which makes for such a multiplicity of clinical symptoms. A general therapy of depression would require us to diagnose and substitute or inhibit each and every cybernetic system which has fallen out of line, and this is a will-o'-the-wisp never to be attained on a practical level.

Fundamentally we should neutralise a "positive" symptom such as anxiety for example, by inhibiting the synthesis of NA or its release, or by blocking noradrenergic receptors. Inhibition of synthesis could also be effected through stimulation of pre-synaptic receptors whose function it is to inhibit the activity of the synthesizing enzyme tyrosine hydroxylase. Reduction in the rate of synthesis offers the possibility of reducing NA-activity, and this improves the symptoms of anxiety and insomnia. Drugs which blocked the receptors for 5-HT also became available, developed with the idea in mind that inhibition of serotonergic activity would improve the inactivity and listlessness of depression.

Let us go back thirty years. The discovery of the antidepressive effects of MAO-inhibitors by Kline, of imipramine by Kuhn and of amitriptyline, initiated a new era of antidepressant treatment.

The use of MAO-inhibitors is based on the fact that they block the breakdown of certain neurotransmitters both within and also outside the neuron, and therefore build up the levels of stored neurotransmitter. This was a perfectly correct therapeutic design, but it did not lead to a decisive success because the MAO-inhibitors available at that time were not specific enough, i.e. they did not produce a specific elevation of NA or of DA or of 5-HT stores. Imipramine and amitriptyline block mainly the re-uptake of biogenic neurotransmitter amines (see Fig. 6). This produces an accumulation of neurotransmitters in the synaptic cleft, which signifies an increase of neurotransmitters in a biologically effective location.

Kielholz has developed a heuristically very useful scheme of classification which is based on the desired effect of the various antidepressive drugs. First of all he distinguishes drugs with a dynamic action, among which he includes tranylcypromine plus trifluoperazine, imipramine, clomipramine and melitracen. The anxiolytics are at the opposite pole, and include amitriptyline (alone or in combination with chlordiazepoxide), doxepine, and mianserin. Although this scheme points us in the right direction, and easy as it may be for the practising physician to understand and use, it can, however, only be successful in its application to treatment if it is adapted to suit individual cases. From a purely practical standpoint the problem of variability can be got round to some extent by prescribing the drugs with the dynamic action in the morning and the ones with a sedative, relaxing effect in the evening. My own preference (WB) is to use tranylcypromine plus trifluoperazine, the most potent stimulant drug, one tablet morning and afternoon, together with selegiline, and then to give amitriptyline (25–50 mg) to settle the patient down in the evening. In the combination of the MAO-inhibitor tranylcypromine with the neuroleptic drug trifluoperazine, the active substance is the tranylcypromine, which blocks the breakdown of the catecholamines and so increases arousal. The small amount of the neuroleptic is to block the postsynaptic DA-receptor. The principle of this treatment is therefore first, that MAO inhibition increases the amounts of the catecholamines in their storage sites, and secondly the mild blockade of postsynaptic DA-receptors. These principles account for the effectiveness of this drug-combination in promoting NA-induced arousal. Any side-effects are usually a sign of excessive dosage. What one generally finds in such circumstances is agitation, inner turmoil and insomnia. What one has to do then is to cut the dose down to half a tablet in the morning, at the same time adding one or two capsules of amitriptyline plus chlordiazepoxide in the evening.

Package inserts always caution against the use of tranylcypromine with tricyclic antidepressants, but this absolute exclusion cannot be justified. I have been prescribing such combinations for over thirty years, and have never encountered any severe side-effects (e.g. acute blood pressure crises) with individually tailored dosages. If a patient becomes more agitated, worried and anxious, one just has to cut back a little on the dosage of tranylcypromine. If the next morning the amitriptyline "hangover" makes the patient too listless, then one has to cut back a little on the evening dose of amitriptyline as well. Nowadays the smooth introduction of the optimal drug regime is no longer a serious clinical problem. A less potent alerting drug is the combination of melitracen with flupenthixol: the first drug is mildly stimulating, the second is a neuroleptic. This combination is particularly useful against masked depressions and can also be used where the tranylcypromine/trifluoperazine combination is not well tolerated.

An antidepressive drug which acts on all the components of a depression is maprotiline. In some patients it improves drive, in others it promotes sleep. Its antidepressive effect is, however, milder than that of amitriptyline. To promote sleep requires a dose of at least 75 mg. Where the symptoms are less severe, less powerful drugs are indicated. Thus melitracen and dibenzipine are among the milder antidepressive drugs, and the milder sedative drugs include mianserin, lofepramine and doxepin.

Patients, who in addition to depression have signs of phobic compulsions (claustrophobia, dislike of being watched) may derive benefit from sulpiride, a mild neuroleptic drug. Its effect can be very variable, however: some patients report that it cheers them up and overcomes their phobic reactions, but also complain about insomnia as side-effect. Other patients do report an effect against their phobias, but complain about lassitude and a sleep-compulsion. However, by suiting the precise nature and dosage of the treatment to the individual patient it is possible to achieve one's therapeutic objectives in almost every case.

There is an unfortunate habit among doctors, found all too often, of senseless administration of tranquilizers. Long-standing anxiety is treated with the whole gamut of benzodiazepines from oxazepam to lorazepam, from bromazepam to chlordiazepoxide. As Pöldinger is constantly stressing, such drugs may be recommended in the initial phase of antidepressive treatment, in combination with genuine antidepressive drugs, only where a sedative effect is urgently needed because of the danger of suicide or because of acute anxiety. The tranquilizers have the advantage of rapid action, while the antidepressive drugs require a certain period before their effect becomes apparent. Tranquilizers have only a minimal effect on the re-establishment of the biochemical balance, and what effect there is is most likely to be via GABA-receptors acting on the NA-systems.

A further bad prescribing habit, and one which is becoming more prevalent, is the administration of neuroleptic drugs in chronic depression to promote sleep: e.g. thioridazine, fluphenazine, zuclopenthixol or melperon at night. Neuroleptics achieve their sedative effect by blocking post-synaptic receptors. This may be indicated in schizophrenic processes, but such intervention in neuronal function never produces a re-establishment of a disturbed biochemical equilibrium. It is of course perfectly easy to sedate an extremely agitated, frightened, or even frenzied patient with an injection of fluphenazine, but in the long term this will do nothing for the depression, except to make the patient easier to manage: it is like a quasi-lobotomy.

Very generally we can conclude, in our present state of knowledge based on research and clinical experience, that a substitution of inadequately available neurotransmitters by their physiological precursors is difficult and not entirely satisfactory. Substitution with precursors can lead to the desired objectives when the dose and the time of its administration have been correctly chosen. However, up to now it has not proved possible to achieve a re-activation of inactive enzymes such as tyrosine hydroxylase, aromatic amino acid decarboxylase, dopamine β-hydroxylase or tryptophan hydroxylase in order to restore a lost ability to synthesise neurotransmitters.

Inhibition of MAO activity does lead to accumulation of biogenic neurotransmitters, and since MAO-inhibitors have become available, their clinical application has led to effective modes of antidepressive treatment. Selegiline inhibits MAO-B and leads to DA-enrichment. Since biochemical analyses demonstrated a deficiency in DA in various brain regions, I have administered selegiline (5–10 mg twice daily) combined with phenylalanine (250 mg daily) orally. In patients with retarded depressions this combination has a dynamic effect which sets in much sooner than with the more usual antidepressive drugs. Patients with severe depressions are started on 10 mg selegiline plus 250 mg phenylalanine daily, and this is continued for two to three weeks. In two thirds of the patients there is a noticeable improvement in their mood, in their enjoyment of life and in interest in their surroundings even after only one week of treatment. When I am satisfied that there is some improvement, I persevere with this treatment. For long-term treatment a single tablet of selegiline (5 mg in the morning) is usually enough. Occasional side-effects include restlessness, agitation and trouble with sleeping, and these can be dealt with by the addition of 10 mg amitriptyline twice a day. Mild sleep disturbances can be handled by giving L-tryptophan, 500 mg at night. Since retarded depressions far exceed the agitated depressions among my patients, my standard oral treatment is as follows:

Drug	Morning	Noon	Night
Parstelin (2–4 tablets)	1	1	0
Phenylalanine (250 mg)	1	1	0
Saroten (10–25 mg)	0	0	1

If there are any adverse reactions the dosage may be cut down. If the mood disturbance recurs and the dejected state of mind persists, I add saroten 25 mg three times a day. Success, as always, depends on a delicate harmonization of the individual active components. If, for example, infusion of selegiline is giving the desired dynamic effects but the patient starts losing sleep, then I add an infusion of 500 mg tryptophan at night.

Since depressive illness does not involve any ongoing process of degeneration, the only way of achieving a genuine increase in neurotransmitter amines is by inhibition of the enzymes that are responsible for their breakdown as well as blockade of re-uptake mechanisms, in contrast to Parkinson's disease, where cell atrophy makes it necessary not only to inhibit MAO but also to add the precursor amino acid.

One substance which the biochemically-oriented clinician does not yet find completely satisfactory is lithium. In my (WB) experience it is very effective against strictly periodic endogenous depressive phases, where it frequently succeeds in suppressing the manic phases. We can assume that there is a connection between lithium and some kind of effect on the neurotransmitters and on the re-establishment of their mutual biochemical balance. Since the results of only relatively few biochemical analyses on postmortem brain material from depressed patients are available, we will have to try in future to apply the modern imaging techniques (e.g. NMR or PET-scanning) to obtain further evidence of disturbed metabolic processes in the brainstem of depressive patients.

Autonomic-affective dysfunctions

As we have repeatedly emphasised, the right balance between the various systems of neurotransmitter amines and neuropeptides is the essential precondition for our normal state of mind, and this balance is maintained by a variety of feedback mechanisms. As a clinical example we might mention the striato-nigral GABA system. The neurotransmitter GABA travels to the substantia nigra via GABA-ergic nerve tracts. There GABA blocks the activity of DA in the nerve cells of the substantia nigra, and therefore the transport of DA to the striatum. If the level of DA is too low, then there is a negative feedback control mechanism which inhibits GABA-ergic activity. The resulting fall in the concentration of GABA in the cells of the substantia nigra increases the availability of DA in the striatum. In Parkinson's disease the stimulation of DA synthesis in the cells of the substantia nigra improves the patient's akinesia. Conversely, too great an activity in the striatum similarly causes a feedback-regulation of GABA-ergic activity. Thus we see that the balance of neurotransmitters in the striatum, maintained by these feedback mechanisms, is the essential basis for the whole normally-occurring range of our involuntary actions. Since the striatum and the substantia nigra also contain other neurotransmitters, our biochemically balanced system does not represent a counterpoised equilibrium, it is more like a global balance, analogous to a galaxy. Various substances stimulate or inhibit the balancing feedback mechanisms. Pain-constricting neuropeptides such as enkephalins and endorphins block the pain-conducting nerve fibres: for example opiates inhibit the release of substance P. The accentuation of pain, just as its inhibition, doubtless exerts the corresponding stimulatory or inhibitory influences on our motor behaviour. If a boxer in a fight receives a painful blow, then this produces a violent reaction and a feedback mechanism releases potentiated aggressive and motor actions.

The whole of our normal behaviour and all our repertoire of motor functions is conditioned by the homeostasis between various agonistic and antagonistic neurotransmitters and neuromodulators. We can expect that inadequate feedback regulation, caused by insufficient stimulation of receptors on the one hand and hypersensitive receptor activity on the other, will produce pathological disturbances. Nowadays we recognise a whole range of substances which stimulate post-synaptic receptors, such as apomorphine, bromocriptine and lisuride, and we are likewise aware of the large group of neuroleptic receptor-blocking drugs.

Insufficient synthesis of a neurotransmitter would be another causal factor, for example, a reduced tyrosine hydroxylase activity that cannot be compensated for by any drug, leading to inadequate DA-synthesis. The clinical symptom of this deficient regulation is an akinesia or an emotional disturbance of arousal.

Finally, a pathological symptom can also be called forth through an increased or accelerated turnover-rate. An increased peripheral turnover is reflected by an increased urinary concentration of the degradative end-product of the neurotransmitter. Thus we see an increased urinary HVA excretion in a patient who is a "fidgety Phil", and a reduced urinary HVA output in anorexia nervosa. In this way the biochemical analysis of precursor amines in the blood and in the CSF and of their metabolites in the urine enables us to obtain evidence of various CNS- or peripherally-conditioned functional disturbances. This allows us to apply limited but rationally designed corrective measures, which will certainly be vastly improved by future advances in our knowledge and by increasing availability of drugs with a specifically designed mode of action. The principle of feedback regulation of biochemical imbalances is equally applicable to the periphery. If someone's blood-sugar falls, then he develops an appetite. There are only two possibilities for dealing with this: either he eats and the resulting increase of blood sugar restores the normal level and feelings of hunger disappear. Alternatively, if external circumstances prevent him getting something to eat, then the hypoglycaemia stimulates receptors which are involved with adrenaline: this promotes the breakdown of glycogen stores and re-establishes the blood glucose levels. The hunger pangs disappear in both cases, but the latter mechanism only keeps hunger at bay for a limited period. When all the autonomic, psychological and affective functions of a person are regulated to maintain optimum conditions through feedback mechanisms, how is it that our involuntary and emotional balance is so frequently disturbed nowadays?

The Austrian researcher Hans Selye put forward the concept of "stress". What he meant by this was the way we react to all our environmental pressures by a series of adaptive processes. At the time he was doing his research he was only able to obtain objective evidence of these processes through measurements of a few parameters, for example the increased activation of the pituitary-adrenal-cortical axis in response to stress. This stress is nothing abnormal, it is only the organism's normal physiological response to any kind of external pressure (inflammations, organic or psychological traumas, information overload by either conscious or sensory i.e. optical or acoustical etc. input). Normal stress is therefore evidence of an optimised adaptation, but when the external pressures exceed the limits of adaptive capability then we begin to see signs of autonomic and emotional disturbance. These may range from insomnia and loss of appetite to irritability and early fatigue, and are the expression of an insufficient capability for adaptation. Someone who has spent the whole day giving his full attention to a job which makes great intellectual or physical demands, let us say a heavy goods vehicle driver, requires an excess of NA-activity in order to maintain the necessary degree of vigilance. The balance between NA and the other neurotransmitters has been displaced in favour of NA, and this prevents him from dropping off to sleep. The required chronic NA-hyperactivity leads to an increase in heart-rate and in blood-pressure, and to loss of appetite etc. If now there should appear a further irritant in his surroundings, let us say that he lives in rooms which provide too little privacy so that he cannot retreat into a quiet corner, then he may exceed the tolerance of his adaptive feedback mechanisms, and these may no longer be able to restore the biochemical balance. This is called the "overstress syndrome".

The first phase of this disturbance is marked by positive symptoms. If one is a little fatigued one may over-react to minor irritations by an aggressive and irritable response. It is only in the subsequent phase that one's reaction is lack of interest and apathy. The first phase of the overstress syndrome exhibits three main effects:

1. There is a reduced threshold to physical and psychological irritation, and one over-reacts with maximum force. Changes in the weather may also cause noticeable responses such as headaches or migraine, tightness of the chest etc. The slightest disagreement provokes a dramatic reaction out of all proportion to its cause.

2. The prolongation of reaction: a patient who is moody or bad tempered at some earlier trivial irritation may go on exhibiting symptoms for a long time because of an inadequate feedback control mechanism. Thus feedings of anger in the morning may persist in the form of a ill temper and poor appetite at lunchtime and insomnia at night.

3. The radiation effect. A sore throat in the morning may develop into an irritant cough and eventually into a tightness of the chest with anxiety, and may be followed by sweaty palms. Such an irritation of the autonomic system may become aggravated and lead to stomach cramps and constipation. Such radiation effects can no longer be calmly coped with via feedback regulation. If such radiation effects repeatedly recur,

then there is in addition the fear that something serious may be wrong. The clinical picture is neither neurologically nor psychologically consistent or appropriate to any particular disorder: periods of agitation, of anxiety, headaches, dizziness, painful indigestion, insomnia – the symptoms follow one another without rhyme or reason.

The various causes are well established: there may be battles over professional status at the workplace, excessive demands at home or at work, personal overinvolvement because of an inability to maintain a proper detachment – at work or at play – and a disregard for the normal day and night rhythm with a reversal of the normal rest and work relationship because of the external demands of one's occupation (e.g. night workers, medical personnel etc.), and finally, sensory input overload such as excessive noise or glaring lighting.

Normally autonomic and emotional functions operate quite smoothly, because smooth transitions are the most economical. Diurnal rhythmical changes take place quite smoothly, and the changes between night and day and between summer and winter are the prototypes for such smooth transitions. Among human beings there are collective or individual pathological processes which Selbach has described as "switch" processes. A faining fit, i.e. a sudden loss of consciousness is such as switch process: switching from consciousness to unconsciousness. An epileptic fit is also a switch process, though of a different kind. Fainting is a sudden loss of the faculties, while the epileptic fit is an explosive maximum discharge of activity. Among humans in a collective sense one can regard the last world war as a switch process on a vast scale – a switch from peace to a more archaic mode of behaviour. The biological meaning of switch processes is that they occur when for some reason it is not possible to achieve a calm regulatory activity in a smooth transition. Although it may be against common sense when the original model of our world is the smooth transition, yet when the pressure becomes unbearable, then the individual – and indeed the collection of individuals – may resort to a switch process, and in principle such processes are supposed to restore the equilibrium.

In the area of the autonomic-emotional functions, such switching appears in the guise of involuntary attacks. Most occur at night, where a patient may have a sudden access of terror, a dread of impending doom, which may last from 15 to 60 minutes. These attacks happen during the REM-phase (dream sleep) and represent a switch process caused by increased NA activity. The first to describe these attacks was the neuro-surgeon Penfield. Clinically they may be seen as a biochemical imbalance, representing a special case of an over-stress-syndrome.

A second form of autonomic dysfunctions are the so-called syncopic attacks (Schulte), involving loss of consciousness of variable duration usually accompanied by a fall in blood-pressure.

Treatment

These phenomena of autonomic dysfunction are the result of excessive demands made on the patient over a long period, and persistent emotional demands may be just as damaging as persistent physical demands. Similarly chronic infections or chronic abuse of noradrenergic stimulants (such as coffee or amphetamines) can lead to an overstimulation with positive symptoms. One can only apply the appropriate therapeutic measures after establishing the underlying causes by an analysis of the patient's motivations. A man who is having marriage problems because his secretary, with whom he is having an affair, is giving him hell and wants to make him divorce his wife and marry her instead, is hardly going to solve his problems by taking tranquilizers. A senior official or managing director of a company, whose work is getting on top of him, has to be educated about the origin of this autonomic dysfunctions. He has to have it explained to him that the has a choice between lowering the sights of his ambitions or suffering a coronary. An Olympic athlete who by intensive training is overdrawn on his autonomic energy supplies and starts to have problems with sleeping, outbreaks of sweating, sudden attacks of agitation etc. has to cut down on his training programme or give up competitive sport for the time being.

Such psycho-therapeutic advice is not going to lead to a sudden restoration of the biochemical balance from one day to the next. The existing symptoms of the imbalance, such as poor sleep, palpitations, cold sweats, anxiety, which were initially occasioned by quite mild irritations, will persist as conditioned reflexes even after removal of the original causes. So the first phase of treatment has to involve the use of tranquilizers, in order to calm the whole autonomic-emotional computer, and to achieve a gradual and progressive damping down of its wild oscillations. The abundance of benzodiazepines (diazepam, bromazepam, lorazepam, oxazepam) in minimal dosage can considerably shorten the path to recovery. Psychotherapy may also help occasionally, and it should concentrate on trying to explain the biochemical rules that govern the balance between the expenditure of energy and the re-establishment of energy stores.

Problems tend to occur particularly in people's achievement phase (roughly between the ages of 25 and 60), but today's demonic urge to achieve is already coming at us as early as puberty. This applies particularly to parents and sports coaches, who wildly overdo the training of young people for spectacular achievements in sport (swimming, skating, gymnastics), so that even amongst these youngsters an overstress syndrome is no longer a rarity. The only difference between young and old is that young people, when put back onto a more sensible life-style, recover from the effects much more rapidly. In autonomic attacks, like the acute attacks of Penfield, the most effective immediate treatment is an intramuscular injection of 10 mg of diazepam, but tranquilizers are less suitable for long term treatment because they often lead to habituation, though not addiction in the psychiatric sense. After withdrawal of these drugs the symptoms of the irritation come back worse than ever. For these reasons I generally prescribe sedative antidepressant drugs such as amitriptyline, doxepin or mianserin (10 mg). Since these never cause habituation or addiction, they can be used over long periods. In a sudden syncopic fit of the Schulte type, I administer no drug at all, because the patient on-coming to has no adverse feelings. For long term management I recommend antidepressives with an NA-stimulating effect (imipramine, 10–25 mg in the morning and dibenzipine 80 mg morning and noon).

Emaciation and obesity

Anorexia nervosa mostly afflicts young 12–18 year-old girls, and they show a lack of appetite, are constipated, sleep badly, get tired easily and are incapable of normal bodily or mental effort. On examination they have a dry skin, complete wasting of the fat depots, cold fingertips and toes, dull hair, brittle nails, slack muscle tone and diminished tendon reflexes. Emotional involvement is reduced and their overall mood is depressed. Psychoanalysis tends to play up psychic dynamic aspects, and it cannot be denied that psychic traumas are capable of producing a biochemical disturbance, but the dramatic insistence that some traumatic event in early childhood has always *got* to be the cause can scarcely be maintained nowadays. The urinary concentrations of neurotransmitter metabolites are strongly reduced, and this points to the cause: a reduced turnover and/or synthesis. The same tendency to emaciation can also be seen in the late forms of Parkinson's disease, but the difference between the puberal patient with anorexia and the Parkinsonian patient is that the latter does not suffer from lack of appetite: quite the reverse, his appetite is increased. However, in both cases there is a steady loss of bodily weight. There is also not a particularly marked tendency to depression in Parkinson's disease. It may be assumed that treatment with L-dopa and dopamimetic drugs is leading to increased DA-activity in what are ultimately the hypothalamic systems of the hunger and satiety centre.

Treatment with DA-agonists (bromocriptine or lisuride) produces a constriction of milk-production but also a reduction of body weight. Stimulation of ergotropic activity leads to a fall in weight and in extreme cases to emaciation.

The treatment of choice is L-tryptophan, 1000 mg three times a day given by mouth, or a single daily infusion of 3000 mg. Since treatment with sedative antidepressant drugs often produces weight gain, pubertal anorexic patients should be treated with tryptophan and amitriptyline. In four cases which were completely refractory, and after every other attempts at treatment had failed, I finally resorted to electroconvulsive treatment, and achieved a realignment of the biochemical balance and with it a weight gain and a cure.

Neuroses

There is hardly a word or a diagnosis which is so often used in medicine as the word "neurosis". Nowadays there is considerable agreement that "neurosis" represents an illness: but exactly what kind of illness is it?

Is a stomach neurosis with the well-known symptom of acid stomach or gastritis a somatic illness? Is a heart neurosis a cardiac disease or a psychological problem? Is asthma a somatic or a psychic disorder? As far as asthma is concerned, chest physicians incline to the view that the constriction of the bronchi and the increased secretion of mucus are somatic correlates of the disease. On the other hand we also know that very often a psychic conflict may be the trigger for an asthmatic attack. To give a commonplace example, the parents go out for the evening and leave the child at home by itself. The child has an asthmatic attack. But asthmatic attacks can also be triggered by changes in the weather, an occurrence stimulating the autonomic system. And finally, there is a multitude of chemical or biological antigens which can produce an attack. Are alterations in their antigenicity within the framework of a depressive illness to be regarded as psychic or somatic symptoms? Neither the one nor the other: such disturbances are particular imbalances between certain neurotransmitters.

What exactly produces that ailment of civilisation, the neurosis? Every neurotic symptom is of endogenous origin: examples are neurotic insomnia or a neurotic impotence, or the classic neurotic symptom, ejaculatio praecox, or neurotic hot flushes, mainly in the face – every one of these is a autonomic-emotional disturbance, i.e. there are symptoms which on the one hand may be experienced in the mind, perhaps as a groundless anxiety, or with an externalised component, such as fear of being struck by lightning, fear of crossing the street, claustrophobia etc., or on the other hand they may be primary autonomic disturbances, such as palpitations, which then produce secondary symptoms like apprehension and breaking out in sweats.

It is immaterial whether one is talking of organic neuroses or marginal neuroses or whether one is blaming traumatic events in early childhood: we are assuming that all neurotic symptoms are the result of an uncontrolled discharge of neurotransmitters. Such a discharge may release noradrenergic symptoms, including anxiety, rising blood pressure, insomnia and so on. An outflow of DA can produce agitation and unrest ("fidgety Phil"), while a release of 5-HT produces flushing, especially of the face, indigestion, frequency of micturition ("irritable bladder") and cat-naps with loss of muscular tone. Each neurotic symptom seems to depend on an uncontrolled release of one or other neurotransmitter, and this produces a loss of biochemical homeostasis. The pathological component of the neurotic complex is manifested by an inability to effect a smooth restoration of the balance via negative feedback mechanisms.

It is characteristic of a neurosis that even a minor fall in blood-pressure can lead to uncertainty, dizziness and fatigue because the noradrenergic recovery fails to occur. The cause of this may be regarded as a genetically determined lowered stimulation threshold, and this then is the immediate cause of the pathological hyper-reactivity. These are the people who start their history with ". . . As a child I was highly-strung and sensitive: the slightest thing would make me cry or frighten me, or make me feel that I was not treated fairly . . ." and so on and so forth. This hypersensitivity is based on the individual's genetic make-up. Similar symptoms can appear in persons of a perfectly normal stable constitution after severe trauma (encephalitis, severe brain damage) or after prolonged stressful overwork. We are not going to argue whether an asthmatic attack is really a subconscious plea for a return to the maternal breast or not, we are simply saying that for example an uncontrolled release of serotonin or of histamine produces a bronchospasm and leads to an increased secretory effort, and that these can be overcome by administration of a noradrenergic drug or by receptor-blockade. Releasing factors may be changes in the weather, particularly low atmospheric pressure, a humid atmosphere or cold. But psychological triggers such as irritation of various receptors in the bronchial mucosa (e.g. by pollen) may also have similar effects. There is evidently a large variety of these releasing factors, whether organic or psychological, but the biochemical reaction is always the same. The inadequate biochemical adaptability of the neurotic organism can be set off by all sorts of adverse stimuli.

Inadequate synthesis and storage of neurotransmitters, or an accelerated rate of breakdown may all contribute to neurotic symptoms in a given situation. Variations in the sensitivity of receptors may in certain cases also be to blame for absent, inadequate or excessive responses.

According to our viewpoint, the neurosis is a pathological reaction to endogenous or exogenous releasing factors, which are caused by genetically determined or otherwise conditioned defects and which cannot be coped with by the normal feedback control mechanisms. Nowadays, with the recognition of the function of the neurotransmitters it is only a matter of time before we can direct our therapeutic measures in such a way that we can normalize the disturbed neuronal activity and get rid of the symptoms the disturbance has produced.

Psychotherapy has its place in the therapeutic armamentarium, but this does not contradict the biological concepts which we have described here. In our view we might regard *"psychotherapy as the mildest form of psychopharmacological treatment, and treatment with a placebo as the mildest form of psychotherapy"*. This viewpoint is supported by recent investigations which have demonstrated the transmission of sound waves to neurotransmitter systems. This indicates a direct connection between speech, speech modulation etc., biochemical conversion and psychotherapy.

Psychopathic disorders

In the neuroses it is the patient who suffers; in the psychopathic disorders the world about him suffers at the hands of the patient. The neurotic patient has a *functional* biochemical imbalance, which pathologically alters and distorts his feelings and his perception of life. The psychopath has a *structural* biochemical lesion, and this causes him to make inadequate adjustments to his surroundings. I was able to study these abnormal psychopathic modes of behaviour because during the last few decades I observed the changed behavioural patterns of patients who had suffered brain injuries in the last war, and the diseased reactions of patients with Parkinson's disease. In both groups the patients showed me an aggressive side to their character which was quite out of keeping with their personalities prior to their injury or illness.

One Parkinsonian patient in my section had aggressive crises, and in one of them she ran out of her room and with her nailfile scratched the paintwork of a consultant's car that was standing in front of the building. When taxed with this she explained that she was driven by an irresistible urge which she had been completely unable to control or prevent. Medical records not infrequently carry reports by nurses and other care-personnel working in the patients' homes concerning aggressive acts by marriage partners. Another of my Parkinsonian patients exhibited a particularly striking aggressive crisis: normally peaceable and well-adjusted, at various times he was seized by an unusual agitation and wandered about aimlessly. Finally he made a homosexual assault on a young fellow patient. The following day he was back to being the most peaceable, helpful, well-adjusted patient.

When, as the result of our biochemical analysis on our Parkinsonian patients, we discovered the deficiencies in biochemical balance among the various neurotransmitters, we were able to interpret psychopathic reactions as being mostly due to biochemical disturbances, and could then manage them with the appropriate drugs. Without knowledge of the patient's previous history one might regard such deviant behaviour as the expression of an inherently psychopathic character. However, this deviant psychopathic behaviour, when it occurs following encephalitis, Parkinson's disease or damage to the brainstem, clearly shows that the sum total of our interpersonal behaviour is guided – or misguided – by the structures of our brainstem.

The excellent German psychiatrist Kurt Schneider tried for decades to classify psychopaths according to their most prominent pathological behaviour patterns. This classification may have been some help on the descriptive level, but goes nowhere at the genetic level. On the genetic level we may argue about a possible inherited defect in the genetic code, or about environmental damage in the widest possible sense or about an episode of encephalitis in early childhood or even a youthful injury to the brainstem. If we include encephalitis in early childhood among the environmentally-caused permanent changes in a patient's character, then we will also have to include virus infections in the pregnant mother, or toxic damage during pregnancy (nicotine, alcohol, tranquilizers). What is remarkable is that relatively often a black sheep (i.e. an individual who deviates from the normal range of values) pops up in even the most normal healthy families.

On the basis of my own clinical experience I incline to the view that psychopathic disease may be caused by a lesion in the region of the brainstem that has occurred at some previous stage of development. What is crucial is the precise developmental stage at which the lesion occurred. A brainstem lesion in a young adult (20–30 years of age) produces mainly negative symptoms, such as a reduction in the patient's mental psychological and somatic initiative and capability. In young children one often sees an unbridled aggressive behaviour, completely uncontrolled and involving no feelings of remorse or any kind of emotional agitation. After all, remorse and regret are the most elementary "feedback" attempts to put things right again. Such psychopathic children tease and even torture animals, as well as younger and weaker children – one might almost say to the point of injury – without showing any kind of emotion. In cases involving this kind of behaviour it seems reasonable to regard the defect as an uninhibited, almost epileptiform release of aggression-producing neurotransmitters.

From a purely hypothetical viewpoint one might propose a fantastic-sounding theory: at present we do not yet know whether the embryological development of the various neurotransmitters (NA, DA, 5-HT etc.) occurs at exactly the same stage of development. However, it is not too far-fetched to assume a dominance of parasympathetic or serotonergic activity vis-à-vis the effects of the dopaminergic and noradrenergic activities at some stage during embryonic development. At this critical point during the maturation of the neuronal systems, alcohol, drug-abuse etc. are apt to cause irreversible distortions, with the result that in later life there will be an inadequate capability on the part of the organism to adapt to biochemical disturbances. We must investigate the essential causes of later psychopathic behaviour which might lie in the prenatal period, because it is during this time that embryo or later the fetus experiences the whole range of pleasure and discomfort. A physiological maturation of neuronal systems and of their higher control mechanisms is at this stage responsible for the error-free evolution of motor, affective and autonomic functions.

Nowadays ultrasound, and even more, fetoscopy, enable us directly and extensively to view the motor functions of the fetus, and we are convinced that further investigations will also demonstrate aberrations of fetal behaviour as the expression of disturbances in neuronal maturation. Up to the present it has unfortunately been the case that in the treatment of psychopathic illness we have only been able to try to correct defects in attitude and behaviour after the damage has already been done.

From a purely clinical viewpoint we can delineate two main basic principles of psychopathic behaviour: first, hyperactive behaviour with increased aggression, increased compulsion to positive action and with uninhibited power-lust; and second, hypersensitive hypoactive behaviour, with reduced drive, both mentally and physically. Such patients are keenly aware of their failure to achieve anything, and are only too ready to blame the world for their own deficiencies.

Both of these groups can easily be accommodated within the framework of our biochemical concept. The hyperactive psychopaths are characterized by a raised catecholaminergic activity, which is however not checked by any inhibiting neuronal control mechanisms. The missing feedback reaction is precisely the pathological element in their psychopathic behaviour. How often in our daily life, when someone has hurt or upset us, do we exclaim "I could murder that man!" But of course we never do, because our biochemical feedback system quickly restores the right balance. It is the *uncontrolled* increase in activity of any neurotransmitter system which leads to the pathological finding or behaviour.

It is our belief that prenatal damage caused by viral infections or by toxic influences may be implicated as causal factors, as evidenced by the noticeable rise in the prevalence of psychopathic children and adolescents. We may include damage to the gametes, leading to genetic lesions, as one of the occasional causes of psychopathic illness, and this could be the explanation for the origin of a small number of hereditary psychopaths.

Modern man seems collectively to have acquired two basic behaviour traits: first, he seems incapable of putting up with frustration and boredom; and second, he seems to have a stronger desire to demand, either from the state or from society, anything and everything that will help him escape any form of frustration and boredom. If we had to choose between the alternatives of making patients pay for their own medicines and drastically cutting down or expectation of the performance demanded of the expectant mother, we would unhesitatingly plump for the latter, as being more likely to be effective.

A distinct type of psychopathic behaviour, *addiction,* deserves special discussion. A good way of understanding these conditions is to look at the classical animal experiment of Olds and Miller. They implanted small electrodes into rat brains and trained the rats to respond to a conditioned reflex. By pressing a lever the rats could activate a short-wave transmitter to induce an electric stimulus in the brain via the implanted electrodes. These were positioned in various areas of the brain, and their stimulation produced sensations of different intensities either of pleasure or of discomfort. When the "pleasure centre" was stimulated, the sensation of pleasure to the animals was so great that they took no notice of food or of sex and just kept on pressing the lever, up to 1000 times a day, until they became severely undernourished: the instincts for survival and for the continuation of the species were completely eliminated. This kind of behaviour is a good model of addictive illness.

An interesting finding has been the discovery that these pleasure centres contain increased concentrations of DA. DA is also the transmitter of the greatest sensation of pleasure, that of orgasm, and during orgasm it is released into the septum pellucidum. In human beings alcoholism represents the greatest percentage of addictions. The biological reaction produced in the CNS by alcohol is without question the increased release of biogenic amines. The immediate effect of alcohol is an emotional tranquilization; since the dawn of history people have drowned their sorrows in drink, and what it also brings is a flushed face, sweaty palms, an increased appetite, sleepiness, a slowing of reflexes, unsteady gait and stupor – all the characteristic signs of alcohol intoxication. Alcohol must surely be the oldest drug. In man's earliest days the priests tried to control it, and ritual commandments doubtless existed to limit its consumption. But it is not the consumption of alcohol in itself that is the psychopathic symptom, but rather the dependence on drink and the compulsion to go on drinking. Alcohol-conditioned aggressive behaviour is probably an excessive and uncontrolled catecholaminergic reaction. Such aggressive behaviour is often directed against the weaker partner: drivers direct their aggressive impulses against cyclists and pedestrians; husbands against wives and children; and in business the victim is the person at the bottom of the pecking order. Of the different victims of alcoholism we would particularly single out the periodic drinker: this periodic form of addiction is always produced by a biochemical imbalance, just as in masked depressions. This type, in my experience, is the only form of alcoholism which can be treated at all successfully.

There are some specific developmental defects such as ventricular septal defects, neural tube defects or urethral defects which can often be surgically corrected after birth, but developmental defects in human behaviour, caused by deficient feedback mechanisms, the very mechanisms which are normally supposed to regulate our behaviour, managing such defects is not nearly so straightforward. The various social arrangements for coping with the afflicted patients are of course admirable, but the proportion of successes is very disappointing. It is an interesting sociological phenomenon that increased affluence does not seem to make people any happier. It seems rather to increase their demands: people never seem to be satisfied! These continually escalating demands eventually mean that the ability of the individual to achieve his desires is overtaxed, and then there may be a breakdown of his biochemically-governed emotional state and well-being.

In a democratic society there are limits to the restrictions that may be imposed. It is not feasible to put every single drug-dealer in jail, or to prohibit the manufacture and retailing of alcohol. However, there is something that we feel might well be worth discussing: here in Vienna we have various "sheltered workshops" where brain-damaged young people can engage in meaningful employment under the guidance of retired master craftsmen. The tempo of these young people's lives is pitched at a significantly lower level than that of normal people, but in spite of big city pressures, inconveniences and long journeys, they nevertheless regularly attend these work-therapy centres. A living community comes into being there, in which mutual support and consideration for one's equally handicapped fellow human being becomes part of everyday life. The question arises whether some such kind of structured community might not also benefit less severely damaged members of society. Of course people would have to join of their own free will, without compulsion, and the degree of individual freedom within such a community would have to be varied according to individual circumstances. Thus for an alcoholic who habitually drinks away his whole week's wages on a Friday night it would perhaps be sufficient to have his wages doled out to him by his wife. And another, for whom drinking was his spare time hobby, his craving for conviviality might need to be satisfied by making him do his drinking in a club, while the kind of drinker who periodically allows himself to yield unresisting to his compulsion to go on a bender would have to live and work in a closed community. Some of these proposals may sound utopian, but the involved physician feels so helpless when faced with the problems of the psychopath, the addict or the hopelessly inadequate.

Any biochemical therapy with the objective of treating psychopathic illness can only be short-term: long-term treatment is impossible, because in the psychopath his harmonious cybernetic balance is fundamentally disturbed, and so his reactions are completely unpredictable.

Neurotransmitters in old age

The proportion of old people in the population is everywhere on the increase, and so there is naturally a corresponding increase in the number of sick people. Old people very rarely suffer from just a *single* illness, and so it is hard for the practising physician to establish priorities among the variety of necessary medicines. Theoretically most old people need drugs to make them sleep, drugs to help their circulation, drugs to prevent joint pain, antidepressant drugs, drugs to prevent memory loss and drugs to combat fatigue.

Biological involution is best understood by comparison with biological evolution. The baby's first main sensory stimuli are the smell of his mother's milk and the feel of her nipple. These stimuli produce the sucking reflex, which is necessary not only for nourishment, it is also the first pleasurable experience. The next step towards mastering the environment is grasping: the infant grasps everything that comes to hand and carries it to his mouth – his first oral action; or he throws it on the ground – his first aggressive act. His sensory world is limited by his reach, and this makes him reach out for the moon that shines in through the window and into the sphere of what lies within his reach. His next developmental leap occurs when he discovers crawling as a way of widening his territory, and this represents an infantile archetype of nomadic life. If the infantile living space is restricted, by putting him into a play-pen, then the toddler will reach up the bars and achieve standing and walking skills, so when horizontal movement is frustrated, he is driven into a new dimension, the vertical. The infant's mastery of such spheres of influence as are bounded by his senses, areas of interaction of the living organism with its surroundings, increases with his advancing development.

The adult's mastery of his environment through his intellectual abilities takes him to the bounds of the world at large and of the microcosm. At the biological zenith – say around the age of fifty – his horizons start to contract, first gradually, then on a more steeply falling curve. First there is a diminished capacity for motor performance: not many fifty-year-olds are going to break any world records. Then in sex-life: both desire and performance decline and there is a lesser need for emotional ties. Real friendships in later life are rarer. And the spiritual horizons shrink as well: the areas of interest, the urge to explore new worlds and new intellectual dimensions gets less. And in the end the senile person regresses to the oral-anal track of the infant. What is it that causes this shrinking of the motor, emotional and intellectual spheres of action?

Our brains are composed of a number of distinct regions: from an evolutionary standpoint the most ancient is the brainstem, which may be regarded as the battery which powers our state of health and behaviour, and above it lies the cerebral cortex, which is our intellectual "thinking-cap". The functions of these brain regions are regulated via the neurotransmitter substances. Dopamine (DA) is the transmitter for all the involuntary autonomic movements, and also for our emotional drive. Noradrenaline (NA) is the transmitter for the whole sympathetic nervous system in the periphery of the organism (blood pressure, action of the heart etc.). In the brainstem NA is responsible for vigilance, i.e. alertness, and the "arousal reaction", our heightened consciousness. Finally, serotonin (5-HT) is the transmitter substance for the whole digestive processes in the periphery, while in the central nervous system it is the specific sleep-prompting substance, as well as having a generally relaxing effect. The neurotransmitter substance of our "thinking-cap", our cerebral cortex, is acetylcholine (ACh). It is also classically the parasympathetic neurotransmitter.

In old age there is a substantial reduction in the brainstem concentrations of most neurotransmitters. The diminished DA leads to a slowing down and a deterioration of co-ordination, as well as to a poorer posture: there is a tendency to stoop, since DA is above all responsible for man's upright stance. NA too is diminished in old age (in the locus caeruleus and other areas of the brain). The consequence of this is a disturbed vigilance, a diminished alertness, an inability to react to changes in the environment with a heightening of consciousness. Another effect of a deficiency of NA is a slowing and inhibition of resolution and decisiveness. Finally, the lack of 5-HT results in a shortened duration of sleep. The occurrence of several interruptions of sleep during the night are particularly characteristic of old age, as is a reduction in REM-sleep, in which dreams take place. These interruptions are initiated by NA as feedback regulation to prevent sleep from becoming too deep. Peripherally the lowered 5-HT is responsible for reduced salivation (a dry mouth), a poorer appetite and digestive problems. Lack of ACh in old age chiefly affects the memory, but intellectual insights and perceptive abilities, and critical faculties and judgement are also affected.

The reason for the diminished neurotransmitter levels must be sought particularly in a reduced activity of the synthesising enzymes: for example tyrosine hydroxylase (synthesising DA and NA), tryptophan hydroxylase (synthesising 5-HT) and choline acetyltransferase (synthesising ACh). This lowering in enzymic activity leads to a lowering of physiological functional ability. But many neuropeptides, substance P for example, are also reduced in old age. On the other hand, in old age there are changes in the way that the receptors respond to stimulation by neurotransmitters, and this leads to a variability in the response.

If any one particular aspect of these declining neurotransmitter systems becomes accentuated then it may give rise to one of a number of well-known syndromes. Parkinson's disease is caused by a specific reduction of DA-levels; endogenous depression by a lowering of DA, NA and ACh; and senile dementia by a progressive deterioration of cholinergic activity leading to a decline of the intellectual capacity of the brain.

In summary, the whole biological capability of the organism is diminished. The old person has less chemical energy available with which to cope with the need to make adjustments. This renders him more susceptible to adverse stimuli, from changes in the weather to infections, because he lacks the necessary resistance. He is more easily shaken out of his emotional equilibrium because the feedback control mechanisms that are supposed to be available to enable him to make the necessary adjustments are not as adequate as they used to be.

Viewed from a therapeutic standpoint, the question arises, if we can treat Parkinson's disease so effectively with L-dopa and the various additional drugs, and if we can neutralise clinical depression so positively with various antidepressive drugs, why cannot we apply the same drugs – though perhaps at a lower dosage – against the same age-related deficiencies?

The answer is that in fact – after many years' personal experience – small doses of L-dopa plus benserazide (62.5 mg) or L-dopa plus carbidopa (50 mg) given from one to three times a day have been found quite effective. The physiological depression of old age, above all the lack of drive, the lack of concentration and the lassitude, can all be successfully treated with tonic-acting antidepressive drugs. A typical schedule might be imipramine (10 mg), and dibenzipine (80 mg) in the morning, and amitriptyline (10 mg), doxepin (25 mg) and maprotiline (50 mg) at night. I (WB) have been giving antidepressive drugs to patients with Parkinson's disease over a period of several decades, without causing addiction or side-effects, unlike the many tranquilizers, hypnotics and pain-killers which *can* lead to habituation. One is not going to get into trouble over giving a single dose of a tranquilizer for an old person who suffers from anxiety over a minor ordeal such as a journey by air, a birthday party, or some kind of public appearance. In old people tranquilizers are not going to lead to addiction and ever-increasing doses, but they can be habit-forming: such patients then get to the point where they cannot fall asleep without lorazepam, bromazepam or oxazepam. The world will not come to an end if an octogenarian regularly takes a tablet of oxazepam at bedtime, that is to say his health will not suffer in any permanent way, but the compulsion to take the pill is a restriction on his freedom of action, and that is why tranquilizers should only be prescribed as a temporary measure.

Basically, old people generally react to medication less rapidly, but excessive doses quickly lead to side-effects or unwanted reactions. They may not react to 2 mg of diazepam, but after 5 mg they may be quite confused and unsteady the following morning and lack their usual degree of alertness all day. So we see that in old people the margin between effectiveness and over-reaction has become smaller.

A much more difficult problem is how to manage presenile loss of memory and shortening of attention-span. Over decades of systematic trial in a geriatric hospital (neurological unit), pyritinol forte (one tablet morning and noon) has proved itself valuable. Both patients and staff report a significant improvement of mental ability. Senile loss of memory is actually a very reasonable consequence of the diminished ability of our cerebral cortex, especially the temporal lobe, to store impressions, and is due to the decreased number of ganglion cells. The stream of sensory impressions continually coming our way can no longer be stored, i.e. stockpiled as biochemical engrams. Only events that are important to life itself are retained. Our memory has the ability to dredge items of information from the depths of our biochemical memory traces to the surface of our consciousness. If the appropriate ganglion cells have perished, then the remembered scenes cannot be brought to mind. Often, however, a period of rest (sleep or relaxation) may allow the engram to be revived, i.e. become a conscious memory again.

The second most valuable drug has proved to be piracetam. It is reported to boost learning ability. Learning ability may not be so important for old people, but what *is* important is the maintenance and the ability of accurate memory recall. Where one sees the real benefit of the drug in improving mental performance is precisely in situations like the slowing of thought processes in Parkinson's disease, or the disturbances in concentration seen in skull or brain injuries or the slowed and impaired mental function of the multi-infarction syndrome. In acute situations we administer piracetam (3 g) in the form of an infusion, and for ambulant patients the same dose in an ampoule as a drink. Subsequently it may be possible to reduce the dose to 400 mg by mouth three times a day. Another drug of similar type is cerebrolysin.

The use of inhibitors of the degradative enzymes MAO-A, MAO-B and COMT leads to an increase in the neurotransmitter levels. Selegiline at a dose-level of 5–10 mg a day blocks the breakdown of DA. Tranylcypromine particularly inhibits the breakdown of NA, but has to be used with the utmost care because it can produce hypertensive crises, which announce their arrival by massive headache, inner agitation and anxiety. An effective and clinically applicable inhibitor of AChE, the enzyme that breaks down ACh, has so far not come to our attention.

At a more basic level, disturbances of cerebral function in old age can also be produced by an impaired cerebral blood flow or by metabolic (enzymatic) defects. In my own experience, the whole range of cerebral blood-flow promoting drugs is of no real value in improving mental performance. What is important, however, is the improvement of the quality of the blood as a transport medium.

Procaine, which has been recommended for decades, is still dismissed too lightly by classically oriented medical men, and in my experience this is mistaken. As long as twenty years ago we were giving novocaine injections (in total 30 intramuscular injections of 5 ml of novanaest-purum 2%) to patients with chronic neurological diseases in my unit. Of course a patient with a hemiplegic disability or a Parkinson's disease patient is not going to walk better after 10–20 injections of nonocaine, but the general biological capability (both motor and emotional) and alertness were decidedly improved. In other words, although procaine cannot do anything about already destroyed ganglion cells, it is quite plausible to suppose that it could act as a stimulus, particularly to the catecholaminergic and cholinergic neurotransmitter systems in the cells that remain, and this stimulation would then be responsible for the general improvement of biological activity. The blocking of degradative enzymes would also lead to a raising of cholinergic functions. Modern drug strategies include the development of AChE-precursors able to cross the blood-brain-barrier, and the synthesis of compounds designed to act differentially with an agonistic or antagonistic effect on receptors of various neurotransmitters, with the idea of operating at this level to re-establish the cybernetic balance among the whole range of neurotransmitters.

For anyone who carries on research into the biosynthesis, turnover and metabolic breakdown of neurotransmitter substances, the theoretical path to tread is clearly marked out: biochemists and pharmacologists will have to convert our clinical suggestions into hard facts. However, in the case of the multi-infarct-conditioned syndromes of old people (multi-infarct dementia, multi-infarct-related Parkinson's disease) which are caused by deficient blood-flow in the brain, no obvious biochemical approaches suggest themselves.

The main thrust of current therapy is in the direction of modifying the physical properties of the circulating blood in the sense of dilution, of reducing its thrombotic defects and in reducing the clumping of the red cells. In addition the tonic effect of digitalis may also be recommended. The geriatric patient with a weak heart needs some kind of digitalis preparation, the hypertensive patient needs hypotensive drugs, and since the structures that maintain posture often hurt because of the wear and tear on the joints, we need some kind of drug to cope with those problems. Eventually the patient may end up with a goodly number of different drugs. The practising physician has the task of making a wise selection among them, especially when there may also be specialists involved, additionally prescribing quite a spectrum of specific drugs against the particular problems for which the individual patients were referred to them. All in all one has to try not so much to prolong life as to give the senior citizen a good quality of life, and the practising physician is generally the person best qualified to perform this service.

The significance of neurotransmitters for human behaviour

The way we function and the way we act depend very much on the cybernetic harmony of the various neurotransmitters in the organism as a whole. We have talked about the "balance within the microcosm of our brain", and put forward the notion that within our central nervous system there exists a permanent rhythmical feedback mechanism which regulates the balance between the specific actions of the neurotransmitters with a dynamic action such as DA and NA on the one hand and of the relaxing neurotransmitters such as 5-HT on the other. This rhythmical relationship is like a sine curve (the Ying-Yang symbol), the dynamic neurotransmitters with a peak of activity at about 5 p.m. smoothly alternating with the neurotransmitters responsible for relaxation having their maximal activity at 4 a.m. This smooth transition between the predominance of the sympathetic neurotransmitters and that of the parasympathetic ones is the essential requirement for an innate harmony in our being and our behaviour. This most vital biological phenomenon of the daily rhythm ensures that a high catecholaminergic activity predominates during the daylight hours, to be followed at night by a corresponding predominance of parasympathetic, including serotonergic activity. 5-HT is the stimulator of many trophic functions, while NA on the other hand is the activator of wakefulness. Each conscious action of the organism requires a waking consciousness (arousal reaction), which is produced by NA from the reticular formation via the locus caeruleus. The diurnal rhythm is also influenced by the light intensity of the surroundings. Increased lighting produces noradrenergic stimulation in the mid-brain, going with a general increase in activity. A fall in lighting leads to 5-HT stimulation. Melatonin, besides exerting its own action, can also be regarded as a mediator of similar functions. This kind of direction of our biological functions by the intensity of the environmental light level naturally enough means that bright sunshine leads to a noradrenergic hyperactivity, which is reflected in the people's improved performance. The boosting of a patient's biochemical capabilities by extremely bright illumination is among the modern methods of treating clinical depression, and it is quite understandable that the resulting increase of catecholaminergic activity can improve depressive symptoms. An opposite example, the light withdrawal that occurs in northerly and southern polar regions shows a decrease of DA and NA during the periods of darkness, going hand in hand with a lowering of the spirits and a damping down of the emotions generally. In fact clinical experience shows that the suicide rate in almost all polar regions in winter is twice as high as in the middle latitudes. On the other hand, the lighting must not become excessive: the people in southern countries (Greeks, Italians, Spaniards) protect themselves against too much sun with appropriately adjusted lifestyles that include ways of avoiding excess of sunlight (Siestas in the afternoon, the use of sunblinds etc.).

The circadian rhythm is without question a deeply engrained biological archetype, impressed upon the "battery" of our brainstem since ancient times, prior to the evolution of the "self". Such a firmly fixed biochemical programme of adjustment cannot be subject to lasting changes without penalty. One could write an essay about the light-dependent characteristic appearance and behaviour of "southerners" and "northerners": less light in the north, less NA, less arousal, increased 5-HT activity. The increased 5-HT of the north leads to a reduction in drive, in vigilance, and to a slowing of motor behaviour: you only have to compare a northern driver with an Italian one! This little journalistic detour is merely meant to show that every general and particular mode of behaviour is directed by neurotransmitters.

Because neurotransmitters are largely, though not exclusively, synthesised and stored in the brainstem, it must be assumed that they have a particularly significant part to play in our whole instinctual behaviour. Instincts correspond to precisely programmed sequences of actions which are triggered by "key" stimuli, and then proceed according to the preprogrammed path, and reach the desired objective or perhaps not. Such archetypal fixed patterns of behaviour can proceed with great economy of effort, but lack variability of consciously directed effort. A frog who flicks his tongue at a fly may get it – or miss it; a hawk that stoops at a mouse can likewise not alter the pre-programmed swoop, just as no plant can refuse to grow towards the light. The precision of neurotransmitter control is thus a condition for animal survival.

Another example is afforded by the "sparking over" kind of behaviour, as described by Niko Tinbergen: two cocks face each other in a combative posture: the excited aggressive attitude makes us fear that any moment there will be a sudden fierce attack. But instead both birds begin to peck the ground and to preen themselves. What, in biochemical terms, is happening here?

The accent on aggressive preparation has a completely noradrenergic and dopaminergic emphasis. If the two opponents are approximately equally matched, however, an actual fight might well be life-threatening, and so in accordance with the survival instinct a sparking-over reaction takes place, and oral satisfaction or sexual arousal may occur, and so the threatening biological situation is biochemically de-fused.

Another example of such biochemical counteraction of behavioural disturbances is given by stereotypic behaviour. Konrad Lorenz described a well-fed starling which flew around the room chasing flies, and when it caught one it would kill it and gulp it down: but it was not actually catching real flies, the whole instinctual sequence was merely stereotypic behaviour. Having been force-fed with mealy-worms beforehand, the bird's hunting-urge had been frustrated, but its unconsumed catecholaminergic capacity could only be compensated by the stereotypic mock-hunt and in this way the balance between motor activity and satisfaction, i.e. between sympathetic and parasympathetic activities was restored.

Any need which is not satisfied through instinctive actions can lead to frustration, and in the animal world such frustrations are managed through stereotypic behaviour or displacement activity. Human beings, however, frequently do not manage failed instinctual actions with stereotypic behaviour. Men and women who may have eaten too richly at dinner-time often do not have any urge to neutralise excessive and unnecessary consumption of rich food by burning it off with exercise such as a brisk country walk: they generally follow the parasympathetic voracious phase with a serotonergic phase in which they sleep it off instead. However, continuous failure to stick to the archetypal rules of biochemical balance is bound eventually to cause disturbances to the harmony of our mental state and our behaviour. Every frustration contributes to a loss of biochemical balance. Common sense or medical assistance can help in coping with the failure to follow one's instincts, and can restore the biologically essential biochemical balance. A politician who is under attack from the media and has slipped into an NA-conditioned stress-reaction, or a managing director whose business problems have shunted him into a state of anxiety and insomnia, they both need to be directed into a parasympathetic phase of positive thought and behaviour. This can be achieved by the initial use of tranquilizers or by administration of neurotransmitter-precursors (Dopa, DOPS or tryptophan). MAO-inhibitors too, insofar as they have the appropriately-directed mode of action, can also help to restore the balance. Of course psychotherapy and behavioural therapy, both of these are able to re-establish the biochemical equilibrium: as an example we may quote autofeedback training which releases the patient from anxiety and tension by a stimulation of parasympathetic activity.

An important principle is that people should be aware that continued failure to live properly is going to disturb their biochemical equilibrium and that this will be followed by a loss of their biological harmony.

A further example from ethology is significant here: every living being that is in mortal danger has the faculty to sham death in order to become unattractive to the predator. The past-master at this is the opossum, hence the phrase "playing possum". What is going on here, biochemically speaking? The animal in its terror is pre-stimulated both noradrenergically and dopaminergically, and it can fight or flee; both of these reactions require the corresponding surge of a movement-activator (DA) and a heightened consciousness (NA), and without them they would be incapable of instant decisions. At a time of great danger there are, however, feedback reactions that produce a great surge of 5-HT, as well as over-stimulation which leads to a block on catecholaminergic activity. The result is a death-like rigidity, a biochemical switch phenomenon similar to syncope in man. Since a gradual transition into such a condition would be too late, this sudden "playing possum" comes instantly to the rescue. And people who find themselves pushed into positions where they become aggravated and angry and where their aggressive feelings cause them heart twinges and a tighness in the chest would do well to activate a dead-still reflex and so evade a dangerous and threatening situation. The man in the street deals with such situations with two ready-made reactions: when things start getting out of hand he will come on strong and speak out: ". . . something has really got to be done about this!" This call to action is then generally followed by a switch to a dead-still reflex: "Ah well . . . can't be helped!" Irritated by his powerlessness in the face of some problem, in order to cope with his increased NA-release he resorts to the safety of the dead-still reflex which prevents him from getting into a mood of frustration by a feedback mechanism that restores the biochemical balance.

Epilogue

The job of the neurologist is to look at behavioural patterns, at changes in behaviour and in the mental state of patients, and to relate these to bodily abnormalities such as strokes, tumours, cysts, or degenerative disorders. The abnormalities are however directly linked to changes in the dynamic equilibrium, in homeostasis and in functional intraneuronal responses. The structural relationships of these major events can be seen by studying Table 3. In our present state of knowledge we can assume that any changes in the regulatory parameters of the nervous system will lead to changes in behaviour of the most varied kinds, depending on the system affected, the changes that have occurred, and the extent and precise location of the disturbances.

If one can exclude direct physical damage as the cause of a significant behavioural alteration, one often has to consider and to question the possibility of social and psychological problems as possible causes. However, it is often quite impossible to establish a causal connection between changes in neuronal processes and these "causative physical factors", and so one just says "it's psychological".

These two different ways of looking at the problems clearly indicate that there is this interaction between mind and body, between "psyche" and "soma". Psychosomatic changes in a patient's psychological state, that is to say somatically-determined changes in the patient's mental state, are therefore causally connected with biochemical regulatory principles. Personality and the mind – body relationship can be altered not only by tumours, cysts and strokes etc., but also by medicines and psycho-active drugs. Here the rather nebulous concept of "personality" on closer inspection turns into a notion which

Table 3. Description of important biochemical concepts for the understanding of higher brain function

Dynamic equilibrium

- Concentration of substrates and products stay constant over a given period, giving the impression of an equilibrium
- All processes tend towards equilibrium but never actually reach equilibrium because of the continuous addition of substrates and removal of products ("open system")
- $\Delta G' =$ negative
- The dynamic equilibrium is referrable to the system

Homeostasis

- If there is an equilibrium between the supply and the consumption of substrate or between the formation and the removal of products, then this is termed homeostasis
- Homeostasis is also referrable to the system

Balance of neurotransmitters

- Neurotransmitter systems are frequently coupled with other systems of the same kind, and thus integrate the functions of different brain regions
  a) directly (e.g. DA with ACh in the nigro-striatal system)
  b) indirectly (more than two systems)
  c) neurotransmitter coupled to a hormone (DA coupled to prolactin; hypothalamus – pituitary system)
- Neuronal fields represent the higher cybernetic integration. The structure of the neuronal network, the modes of interconnection and the plasticity of the brain are all brought into play to maintain the overall balance

encompasses the patient's whole individuality, and in our view it depends on the individual organisation of neuronal networks including neuronal membrane structures, and consequently on the resulting specific cybernetic properties. The distinction between "strong" and "weak" personalities may be reduced to differences in the effectiveness of interneuronally-occurring functions and how well they can be modulated. Effectively organised feedback mechanisms are essential biochemical and behaviour-conditioning features. *Inadequate biochemical feedback leads to inadequate behavioural feedback*. The sooner a biochemical functional defect can be compensated, the sooner behaviour returns to normal. If a situation swings in a certain direction then there must be a correspondingly increased biochemical swing the other way to even things out. If such a feedback mechanism is lacking, then the situation threatens to get out of hand. Statistically speaking the deviations from the normal distribution of a particular system balance each other out, but extreme deviations represent real emergencies.

Fig. 14 is a diagrammatic representation of the major higher regulatory mechanisms, an attempt at describing the biochemical regulation of behaviour. We are assuming that four transmitter systems are coupled to one another (it could equally well be two, three, or five or even more). They are thus intraneuronally interdependent in their individual functions. The circle drawn through the middle of systems 1–4 represents the determined balance of the system as a whole. Every system is capable of reducing or increasing its activity, according to the requirements (upper part of Figure 14), and we are further postulating that this particular system is responsible for arousal. Normally the system is in a state of dynamic equilibrium, and all its components are in a state of homeostasis or balance. If now some stress is applied to component system 1 (say a biological regulation to a higher functional output), then there is a reinforced induction of system 2, and a intraneuronal feedback mechanism sees to it that the applied stimulus is reduced. In normal circumstances, after these inducing effects have gone round the cycle, after the 4th system the intraneuronal feedback regulation will have resulted in a normalisation of the initially increased activity of the first system. Let us take an example: a 100 meter sprinter is crouched in his starting-blocks, and the starting-pistol goes off – induction of system 1 – the runner converts this induced activity into physical motion

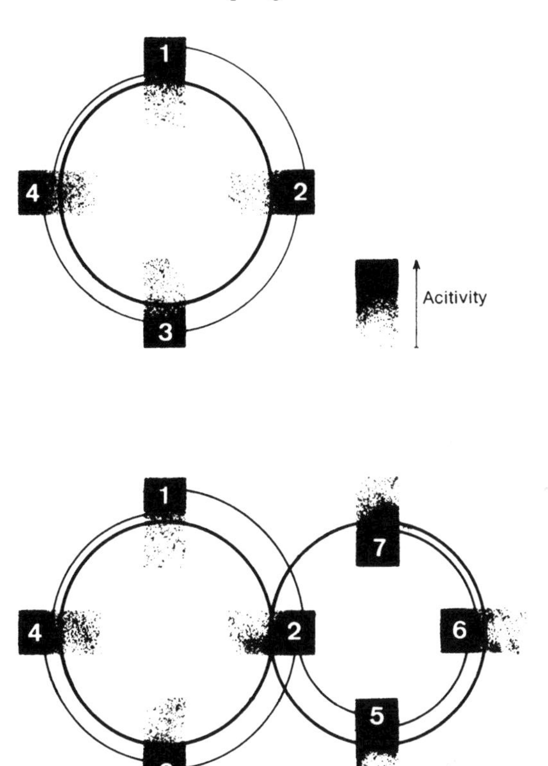

Fig. 14. *Upper:* Simplified hypothetical model of the combined operation of functionally coupled transmitter systems 1–4. Each system can regulate itself, but also modulates the subsequent one. In the normal state the activity of each system has a characteristic value, which is subject to adjustment. Changes induced in any of the systems – say in system 1 – are dealt with via intra- and inter-neuronally linked feedback mechanisms in order to restore the original physiological levels. *Lower:* Hypothetical representation of how two such systems might be coupled together. Systems 1–4 might for example represent "initiative", while systems 2, 5–7 might represent "emotions". The distinct components corresponding to these behavioural states are combined in system 2, and therefore influence one another. The principle underlying the operation of the feedback mechanisms is the same as in the upper diagram.

and speeds for the tape. This reduces the tension and the overall systems down-regulate system 2 back to normality.

The lower part of Fig. 14 again shows the "arousal" system with transmitters 1–4. The second component system is here however also connected to another higher regulatory cycle, which for example might be called "emotion" – systems 2, 5, 6 and 7. Emotion too is subordinated to feedback regulation, as previously described, but emotion and arousal are often intimately interlinked. Thus joy is always closely connected with arousal: the winner throws up his arms in delight, he jumps excitedly up and down, or he may run a lap of honour.

We have no wish to over-complicate these hypotheses by linking in any further systems, but we do want to emphasise the close interaction between mind and body, and to underpin this concept with a biochemical model. It is immaterial here whether the stimulus for a biochemical adjustment is a tumour, a stroke or a cyst; a drug or the spoken word: any of the latter stimuli can be relayed by wave-mechanical coupling to neuronal membranes and through them to neurotransmitter systems. Psychopharmacological drugs, autofeedback training, psychoanalysis, group therapy, religion, yoga and so on, are accordingly all valid routes towards the possible biochemical adjustments. However, in the end we are bound to admit that there is considerable variability in the degree of influence which these various means can bring to bear on the neuronal systems, so not every defect in behaviour can be improved by these forms of treatment or means of regulating our lives.

Reading list

Birkmayer W, Winkler W (1951) Klinik und Therapie der vegetativen Funktionsstörungen. Springer, Wien

Birkmayer W (1951) Hirnverletzungen, Springer, Wien

Birkmayer W (1970) Urbane Anthropologie. Monatskurse ärztl Fortb 20:505–507

Birkmayer W, Danielczyk W, Neumayer E, Riederer P (1972) The balance of biogenic amines as condition for normal behaviour. J Neural Transm 33:163–187

Birkmayer W, Riederer P (1985) Die Parkinson-Krankheit. Biochemie, Klinik, Therapie, 2nd edn. Springer, Wien New York

Birkmayer W (1986) Der Mensch zwischen Harmonie und Chaos, 5th edn. Deutscher Ärtzeverlag

Böhme W (Hrsg) (1980) Herrenalber Texte 23: Wie entsteht der Geist. Gebr Tron KG

Changeux JP (1984) Der neuronale Mensch. Rowohlt

Cohen G (1977) The psychology of cognition. Academic Press

Dale H (1953) Adventures in physiology. Pergamon Press

Ditfurth H von (1976) Der Geist fiel nicht vom Himmel. Hoffmann und Campe

Eccles J (1964) The physiology of synapses. Springer, Berlin Heidelberg New York

Edelman G (1981) Group selection as the basis for higher brain function. In: Schmitt F (ed) The organisation of the cerebral cortex. MIT Press, pp 535–563

Fodor J (1975) The language of thought. Hassocks, Harvester

Fodor J (1981) The mind-body problem. Scient Amer 244/1:114–123

Freud S (1950ff) Gesammelte Werke. Imago

Hebb D (1980) Essay on mind. Lawrence Erlbaum

Henry J, Ely D (1976) Biological correlates of psychosomatic illness. In: Grenell R, Galey S (eds) Biological fondations of psychiatry. Raven Press, pp 945–981

Jovanovic UJ (1976) Schlaf und vegetatives Nervensystem. In: Sturm A, Birkmayer W (eds) Klinische Pathologie des vegetativen Nervensystems. G Fischer, p 363

Kandel E (1976) Cellular basis of behavior. An introduction to behavioral neurobiology. Freeman

Kraepelin E (1896–1915) Psychiatrie. Abel
Lorenz K (1978) Vergleichende Verhaltensforschung. Grundlagen der Ethologie. Springer, Wien New York
Lorenz K (1981) Über tierisches und menschliches Verhalten. Aus dem Werdegang der Verhaltenslehre. Piper
Popper KR, Eccles JC (1982) Das Ich und sein Gehirn. Piper
Vollmer G (1975) Evolutionäre Erkenntnistheorie. Hirzel

Subject index

Made in the USA
Middletown, DE
14 March 2022